I0488406

Status and Understanding of Groundwater Quality in the San Francisco Bay Groundwater Basins, 2007: California GAMA Priority Basin Project

By Mary C. Parsons, Justin T. Kulongoski, and Kenneth Belitz

A product of the California Groundwater Ambient Monitoring and Assessment (GAMA) Program

Prepared in cooperation with the California State Water Resources Control Board

Scientific Investigations Report 2012–5248

U.S. Department of the Interior
U.S. Geological Survey

U.S. Department of the Interior
KEN SALAZAR, Secretary

U.S. Geological Survey
Suzette M. Kimball, Acting Director

U.S. Geological Survey, Reston, Virginia: 2013

For more information on the USGS—the Federal source for science about the Earth, its natural and living resources, natural hazards, and the environment, visit http://www.usgs.gov or call 1–888–ASK–USGS.

For an overview of USGS information products, including maps, imagery, and publications, visit http://www.usgs.gov/pubprod

To order this and other USGS information products, visit http://store.usgs.gov

Suggested citation:
Parsons, M.C., Kulongoski, J.T., and Belitz, Kenneth, 2013, Status and understanding of groundwater quality in the San Francisco Bay groundwater basins, 2007—California GAMA Priority Basin Project: U.S. Geological Survey Scientific Investigations Report 2012–5248, 76 p.

Contents

Contents—Continued

Figures

Figures—Continued

Figure 13. Maps showing concentrations of selected inorganic constituents for USGS-grid and USGS-understanding wells, April–June 2007, and from the California Department of Public Health database for the period April 1, 2004–March 31, 2007, San Francisco Bay study unit, California GAMA Priority Basin Project: nitrate, total dissolved solids, chloride, iron, and manganese 30

Figure 14. Plots showing nitrate, as nitrogen, concentrations related to: classifications of groundwater age, dissolved oxygen, well type, and classification of groundwater age, and depth to top-of-perforations, well type, and land use in USGS-grid and USGS-understanding wells sampled for the San Francisco Bay study unit, California GAMA Priority Basin Project, April–June 2007 34

Figure 15. Graph of total total dissolved solids concentrations relative to depth to top-of-perforation, well type, and geochemical classifications in USGS-grid and USGS-understanding wells in the San Francisco Bay study unit, California GAMA Priority Basin Project, April–June 2007 37

Figure 16. Graph showing detection frequency and maximum relative-concentrations of organic and special-interest constituents detected in USGS-grid wells in the San Francisco Bay study unit, California GAMA Priority Basin Project, April–June 2007 39

Figure 17. Graph showing the detection frequency and maximum relative-concentrations of selected organic and special-interest constituents in USGS-grid wells in the San Francisco Bay Study Unit, California GAMA Priority Basin Project, April–June 2007 40

Figure 18. Map showing concentrations of selected organic and special-interest constituents detected in USGS-grid and USGS-understanding wells for April–June 2007, San Francisco Bay study unit, California GAMA Priority Basin Project: chloroform, solvents, atrazine, and perchlorate 41

Figure 19. Graph showing concentrations of CFC-113 relative to well depth and groundwater age classification and geochemical classification for the San Francisco Bay study unit, California GAMA Priority Basin Project, April–June 2007 46

Figure 20. Graph showing concentrations of perchlorate relative to groundwater age classification for the San Francisco Bay study unit, California GAMA Priority Basin Project, April–June 2007 48

Figure 21. Graph showing concentrations of perchlorate relative to dissolved oxygen concentrations, land-use classification, and groundwater age classification for the San Francisco Bay study unit, California GAMA Priority Basin Project, April–June 2007 48

Figure 22. Predicted probability of detecting perchlorate in groundwater as a function of aridity index and Anthropogenic Score and observed detection frequency and average aridity index in groups of samples for perchlorate concentrations greater than or equal to 0.5 microgram per liter (µg/L) and 1.0 µg/L, San Francisco Bay study unit, California GAMA Priority Basin Project ... 49

Tables

Conversion Factors, Datums, and Abbreviations and Acronyms

Conversion Factors

Inch/Pound to SI

Multiply	By	To obtain
Length		
inch (in.)	2.54	centimeter (cm)
inch (in.)	25.4	millimeter (mm)
foot (ft)	0.3048	meter (m)
mile (mi)	1.609	kilometer (km)
Area		
square foot (ft^2)	0.09290	square meter (m^2)
square mile (mi^2)	2.590	square kilometer (km^2)
Flow rate		
inch per year (in/yr)	25.4	millimeter per year (mm/yr)
Radioactivity		
picocurie per liter (pCi/L)	0.037	becquerel per liter (Bq/L)

Temperature in degrees Celsius (°C) may be converted to degrees Fahrenheit (°F) as follows:

$$°F = (1.8 \times °C) + 32.$$

Temperature in degrees Fahrenheit (°F) may be converted to degrees Celsius (°C) as follows:

$$°C = (°F - 32)/1.8.$$

Specific conductance is given in microsiemens per centimeter at 25 degrees Celsius (µS/cm at 25 °C).

Concentrations of chemical constituents in water are given in milligrams per liter (mg/L) or micrograms per liter (µg/L). One milligram per liter is equivalent to 1 part per million (ppm); 1 microgram per liter is equivalent to 1 part per billion (ppb). Concentrations or activities for radioactive constituents in water are given in picocuries per liter (pCi/L), percent modern carbon (pmc), or tritium units (TU).

Datums

Vertical coordinate information is referenced to the North American Vertical Datum of 1988 (NAVD 88).

Horizontal coordinate information is referenced to the North American Datum of 1983 (NAD 83).

Conversion Factors, Datums, and Abbreviations and Acronyms—Continued

Abbreviations and Acronyms

AL-US	U.S. Environmental Protection Agency action level
E	estimated or having a higher degree of uncertainty
GAMA	Groundwater Ambient Monitoring and Assessment Program
HAL-US	U.S. Environmental Protection Agency lifetime health advisory level
HBSL	health-based screening level
LRL	laboratory reporting level
LSD	land-surface datum
LT-MDL	long-term method detection level
MCL-CA	California Department of Public Health maximum contaminant level
MCL-US	U.S. Environmental Protection Agency maximum contaminant level
MDL	method detection level
MRL	minimum reporting level
nc	no significant correlation
NL-CA	California Department of Public Health notification level
NWIS	National Water Information System (USGS)
PSW	public-supply well
QC	quality control
RSD5-US	U.S. Environmental Protection Agency risk-specific dose at a risk factor of 10^{-5}
SFBAY	San Francisco Bay study unit
SMCL-CA	California Department of Public Health secondary maximum contaminant level
SMCL-US	U.S. Environmental Protection Agency secondary maximum contaminant level
TEAP	terminal electron acceptor process
TT-US	U.S. Environmental Protection Agency treatment technique

Organizations

CDPH	California Department of Public Health (Department of Health Services prior to July 1, 2007)
CDPR	California Department of Pesticide Regulation
CDWR	California Department of Water Resources
LLNL	Lawrence Livermore National Laboratory
NAWQA	National Water-Quality Assessment Program (USGS)
NWQL	National Water Quality Laboratory (USGS)
SWRCB	State Water Resources Control Board (California)
USEPA	U.S. Environmental Protection Agency
USGS	U.S. Geological Survey

Conversion Factors, Datums, and Abbreviations and Acronyms—Continued

Selected chemical names

Ammonia-N	ammonia as nitrogen
^{14}C	carbon-14
$CaCO_3$	calcium carbonate
CFC-113	1,1,2-trichlorotrifluoroethane
DO	dissolved oxygen
^3H	tritium
MTBE	Methyl *tert*-butyl ether
NDMA	*N*-nitrosodimethylamine
Nitrate-N	nitrate as nitrogen
Nitrite-N	nitrite as nitrogen
PCE	tetrachloroethene
TCA	1,1,1-trichloroethane
TCE	trichloroethene
TDS	total dissolved solids
THM	trihalomethane
VOC	volatile organic compound

Status and Understanding of Groundwater Quality in the San Francisco Bay Groundwater Basins, 2007: California GAMA Priority Basin Project

By Mary C. Parsons, Justin T. Kulongoski, and Kenneth Belitz

Abstract

Groundwater quality in the approximately 620-square-mile (1,600-square-kilometer) San Francisco Bay study unit was investigated as part of the Priority Basin Project of the Groundwater Ambient Monitoring and Assessment (GAMA) Program. The study unit is located in the Southern Coast Ranges of California, in San Francisco, San Mateo, Santa Clara, Alameda, and Contra Costa Counties. The GAMA Priority Basin Project is being conducted by the California State Water Resources Control Board in collaboration with the U.S. Geological Survey (USGS) and the Lawrence Livermore National Laboratory.

The GAMA San Francisco Bay study was designed to provide a spatially unbiased assessment of the quality of untreated groundwater within the primary aquifer system, as well as a statistically consistent basis for comparing water quality throughout the State. The assessment is based on water-quality and ancillary data collected by the USGS from 79 wells in 2007 and is supplemented with water-quality data from the California Department of Public Health (CDPH) database. The primary aquifer system is defined by the depth interval of the wells listed in the CDPH database for the San Francisco Bay study unit. The quality of groundwater in shallower or deeper water-bearing zones may differ from that in the primary aquifer system; shallower groundwater may be more vulnerable to surficial contamination.

The first component of this study, the status of the current quality of the groundwater resource, was assessed by using data from samples analyzed for volatile organic compounds (VOCs), pesticides, and naturally occurring inorganic constituents, such as major ions and trace elements. Water-quality data from the CDPH database also were incorporated for this assessment. This status assessment is intended to characterize the quality of groundwater resources within the primary aquifer system of the San Francisco Bay study unit, not the treated drinking water delivered to consumers by water purveyors.

Relative-concentrations (sample concentration divided by the benchmark concentration) were used for evaluating groundwater quality for those constituents that have Federal and (or) California benchmarks. A relative-concentration greater than (>) 1.0 indicates a concentration greater than a benchmark, and a relative-concentration less than or equal to (≤) 1.0 indicates a concentration equal to or less than a benchmark. Relative-concentrations of organic and special-interest constituents were classified as *low* (relative-concentration ≤ 0.1), *moderate* (0.1 < relative-concentration ≤ 1.0), or *high* (relative-concentration > 1.0). Inorganic constituent relative-concentrations were classified as *low* (relative-concentration ≤ 0.5), *moderate* (0.5 < relative-concentration ≤ 1.0), or *high* (relative-concentration > 1.0). A lower threshold value of relative-concentration was used to distinguish between low and moderate values of organic constituents because organic constituents are generally less prevalent and have smaller relative-concentrations than naturally occurring inorganic constituents.

Aquifer-scale proportion was used as the metric for evaluating regional-scale groundwater quality. High aquifer-scale proportion is defined as the percentage of the primary aquifer system that has relative-concentration greater than 1.0 for a particular constituent or class of constituents; proportion is based on an areal rather than a volumetric basis. Moderate and low aquifer-scale proportions were defined as the percentages of the primary aquifer system that have moderate and low relative-concentrations, respectively. Two statistical approaches—grid-based and spatially weighted—were used to evaluate aquifer-scale proportion for individual constituents and classes of constituents. Grid-based and spatially weighted estimates were comparable in the San Francisco Bay study unit (90-percent confidence intervals).

Inorganic constituents with health-based benchmarks were present at high relative-concentrations in 5.1 percent of the primary aquifer system, and at moderate relative-concentrations in 25 percent. The high aquifer-scale proportion of inorganic constituents primarily reflected high aquifer-scale proportions of barium (3.0 percent) and nitrate (2.1 percent). Inorganic constituents with secondary maximum contaminant levels were present at high relative-concentrations in

14 percent of the primary aquifer system and at moderate relative-concentrations in 33 percent. The constituents present at high relative-concentrations included total dissolved solids (7.0 percent), chloride (6.1 percent), manganese (12 percent), and iron (3.0 percent).

Organic constituents with health-based benchmarks were present at high relative-concentrations in 0.6 percent and at moderate relative-concentrations in 12 percent of the primary aquifer system. Of the 202 organic constituents analyzed for, 32 were detected. Three organic constituents were frequently detected (in 10 percent or more of samples): the trihalomethane chloroform, the solvent 1,1,1-trichloroethane and the refrigerant 1,1,2-trichlorotrifluoroethane. One special-interest constituent, perchlorate, was detected at moderate relative-concentrations in 42 percent of the primary aquifer system.

The second component of this work, the understanding assessment, identified some of the primary natural and human factors that may affect groundwater quality by evaluating land use, physical characteristics of the wells, and geochemical conditions of the aquifer. Results from these evaluations were used to explain the occurrence and distribution of constituents in the study unit.

Introduction

To assess the quality of ambient groundwater in aquifers used for drinking-water supply and to establish a baseline groundwater-quality monitoring program, the California State Water Resources Control Board (SWRCB), in collaboration with the U.S. Geological Survey (USGS) and Lawrence Livermore National Laboratory (LLNL), implemented the Groundwater Ambient Monitoring and Assessment (GAMA) Program (website at: *http://www.waterboards.ca.gov/gama*). The statewide GAMA Program currently consists of four projects: (1) the GAMA Priority Basin Project, conducted by the USGS (website at: http://ca.water.usgs.gov/gama/); (2) the GAMA Domestic Well Project, conducted by the SWRCB; (3) GAMA Special Studies, conducted by LLNL, and (4) the online database GeoTracker GAMA, conducted by the SWRCB. On a statewide basis, the Priority Basin Project focused primarily on the deeper part of the groundwater resource, and the SWRCB Domestic Well Project generally focused on the shallower aquifer systems. The primary aquifer system may be at less risk of contamination than the shallow wells, such as private domestic and environmental monitoring wells, which are closer to surficial sources of contamination.

As a result, concentrations of contaminants, such as volatile organic compounds (VOCs) and nitrate, in shallow wells can be greater than in wells screened in the deeper primary aquifer (Kulongoski and others, 2010; Landon and others, 2010).

The SWRCB initiated the GAMA Program in 2000 in response to a legislative mandate (State of California, 1999; 2001a). The GAMA Priority Basin Project was initiated in response to the Groundwater Quality Monitoring Act of 2001 to assess and monitor the quality of groundwater in California (State of California, 2001b). The GAMA Priority Basin Project is a comprehensive assessment of statewide groundwater quality designed to improve understanding and identification of risks to groundwater resources and to increase the availability of information about groundwater quality to the public. For the Priority Basin Project, the USGS, in collaboration with the SWRCB, developed monitoring plans to assess groundwater basins through direct sampling of groundwater and other statistically reliable sampling approaches (Belitz and others, 2003; California State Water Resources Control Board, 2003). Additional partners in the GAMA Priority Basin Project include the California Department of Public Health (CDPH), the California Department of Pesticide Regulation (CDPR), the California Department of Water Resources (CDWR), and local water agencies and well owners (Kulongoski and Belitz, 2004).

The ranges of hydrologic, geologic, and climatic conditions in California were considered in this program's assessment of groundwater quality. Belitz and others (2003) partitioned the State into 10 hydrogeologic provinces, each with distinctive hydrologic, geologic, and climatic characteristics (fig. 1). All of these hydrogeologic provinces contain groundwater basins and subbasins designated by the CDWR (California Department of Water Resources, 2003). Groundwater basins generally consist of relatively permeable, unconsolidated deposits of alluvial or volcanic origin. Eighty percent of the approximately 16,000 public-supply wells in California are in designated groundwater basins. Groundwater basins and subbasins were prioritized for sampling on the basis of the number of public-supply wells, with secondary consideration given to municipal groundwater use, agricultural pumping, the number of historically leaking underground fuel tanks, and registered pesticide applications (Belitz and others, 2003). The 116 priority basins, and additional areas outside defined groundwater basins, were grouped into 35 study units, which include approximately 95 percent of the public-supply wells in California. The San Francisco Bay study unit is composed of eight groundwater basins in the Southern Coast Ranges hydrogeologic province.

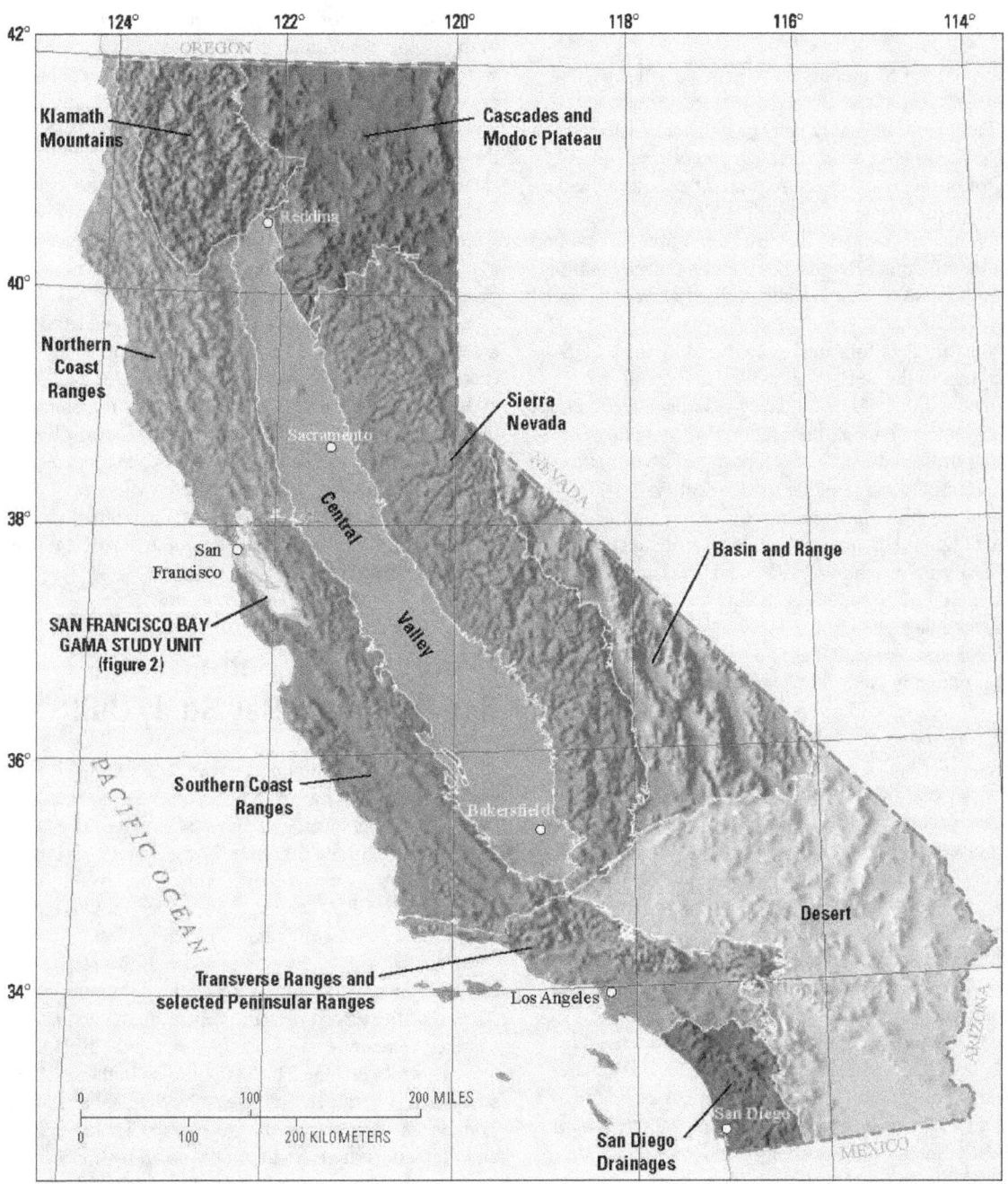

Shaded relief derived from U.S. Geological Survey
National Elevation Dataset, 2006.
Albers Equal Area Conic Projection

Provinces from Belitz and others, 2003.

Figure 1. San Francisco Bay study unit, California GAMA Priority Basin Project, and the California hydrogeologic provinces.

Purpose and Scope

The purposes of this report are to provide a (1) *study unit description*: description of the hydrogeologic setting of the San Francisco Bay study unit (fig. 1), hereinafter referred to as the SFBAY study unit, (2) *status assessment*: assessment of the status of the current quality of groundwater in the primary aquifer system in the SFBAY study unit, and (3) *understanding assessment*: identification of the natural and human factors affecting groundwater quality and explanation of the relations between water quality and selected potential explanatory factors.

Water-quality data for samples collected by the USGS-GAMA Program in the SFBAY study unit and details of sample collection, analysis, and quality-assurance procedures for the SFBAY study unit are reported in Ray and others (2009). Using those same data, this report describes methods used in designing the sampling network, identifying CDPH data for use in the status assessment, estimating aquifer-scale proportions of relative-concentrations, analyzing ancillary datasets, classifying groundwater age, and assessing the status and understanding of groundwater quality by statistical and graphical approaches.

The status assessment includes analyses of water-quality data for 43 production wells selected by the USGS for spatial coverage of 1 well per grid cell across the SFBAY study unit (hereinafter referred to as USGS-grid wells). Most of these wells were public-supply wells, but a few other types of wells with similar perforation depth intervals to the USGS-grid wells also were sampled. Samples were collected for analysis of organic constituents, such as VOCs and pesticides, and inorganic constituents, such as major ions and trace elements. Water-quality data from the CDPH database also were used to supplement data collected by USGS for the GAMA Program. The resulting set of water-quality data from USGS-grid wells and selected CDPH wells was considered to be representative of the water quality in the primary aquifer system in the SFBAY study unit; the primary aquifer system is defined by the depth intervals of the wells listed in the CDPH database for the SFBAY study unit. USGS-GAMA status assessments are designed to provide a statistically robust characterization of groundwater quality in the primary aquifer system at the basin-scale (Belitz and others, 2003). The statistically robust design also allows basins to be compared and results to be synthesized at regional, statewide, and national scales.

To provide context, the water-quality data discussed in this report were compared to State and Federal drinking-water benchmarks, both regulatory and non-regulatory. The assessments in this report characterize the quality of untreated groundwater resources in the primary aquifer system within the study unit, not the drinking water delivered to consumers by water purveyors. After withdrawal from the ground, water typically is treated, disinfected, and (or) blended with other waters to maintain acceptable water quality. Benchmarks apply to water that is served to the consumer, not to untreated groundwater.

The understanding assessment uses data from 36 wells sampled by the USGS for the purpose of understanding (hereinafter referred to as USGS-understanding wells), in addition to data from the 43 USGS-grid wells sampled for the status assessment, to determine the relations between water quality and selected potential explanatory factors and to identify the natural and human factors affecting groundwater quality. Potential explanatory factors examined included land use, depth of well, indicators of groundwater age, and geochemical conditions.

Hydrogeologic Setting of the San Francisco Bay Study Unit

The SFBAY study unit covers approximately 620 square miles (mi^2) in San Francisco, San Mateo, Santa Clara, Alameda, and Contra Costa Counties. The SFBAY study unit lies in the Southern Coast Ranges hydrogeologic province (fig. 1), and contains eight groundwater basins defined by CDWR: Marina, Lobos, Downtown, Islais Valley, South San Francisco, Visitacion Valley, Westside, and Santa Clara Valley (fig. 2). For the purpose of this study, these eight groundwater basins were grouped into one study unit. As part of the Priority Basin Project, untreated groundwater samples were collected from 79 wells in the SFBAY study unit from April 23 to June 21, 2007 (Ray and others, 2009). This study unit is bounded on the east by the Diablo Range and Franciscan complex, on the west by the Santa Cruz Mountains and the San Andreas Fault, on the north by the Golden Gate Strait, and on the south by the Santa Clara Valley groundwater divide (figs. 2, 3). This groundwater divide at Cochrane Road in Morgan Hill separates the northerly flow of water toward San Francisco Bay from the southerly flow of water toward Monterey Bay (Moran and others, 2002).

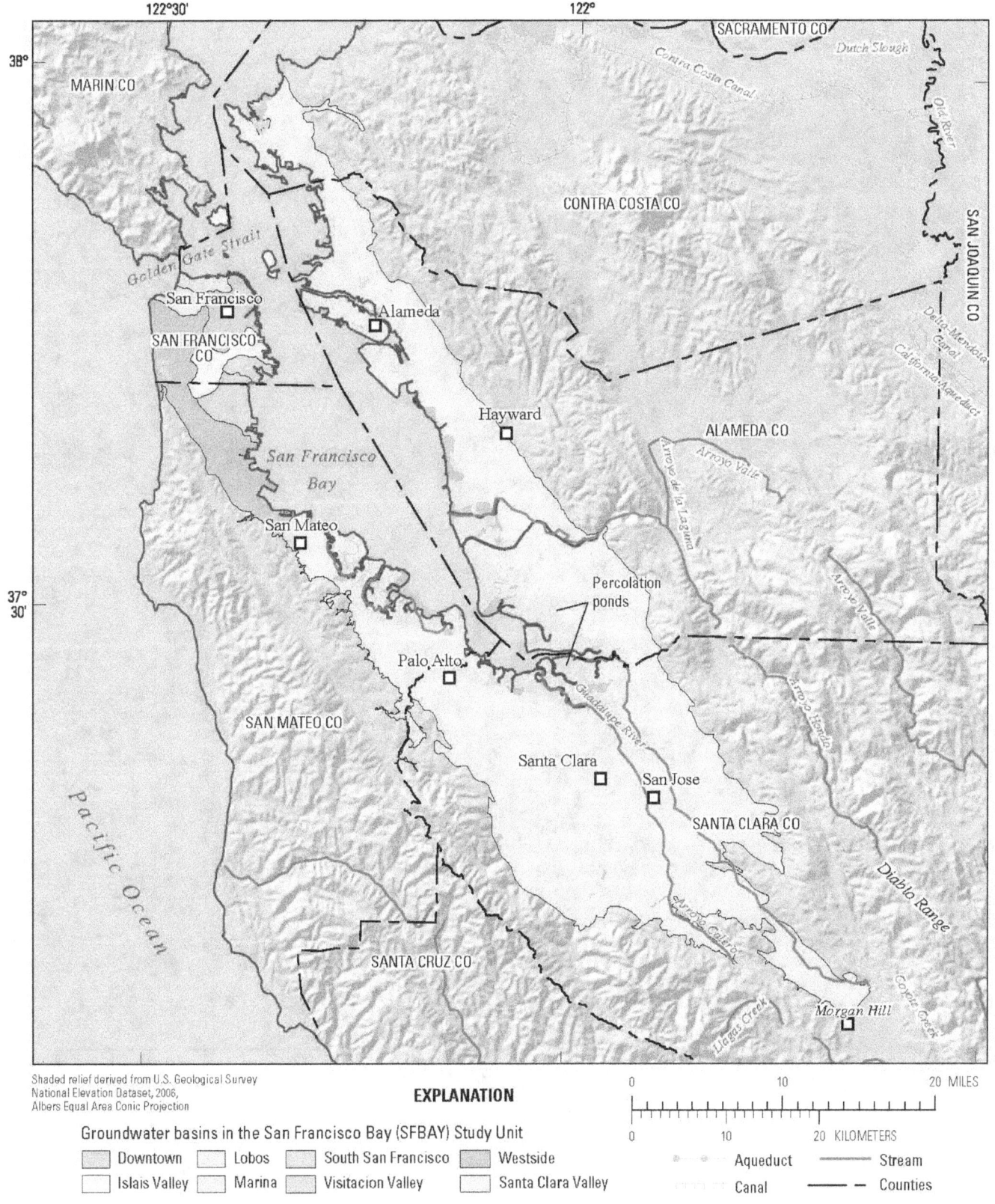

Figure 2.   Geographic features of the San Francisco Bay study unit, California GAMA Priority Basin Project.

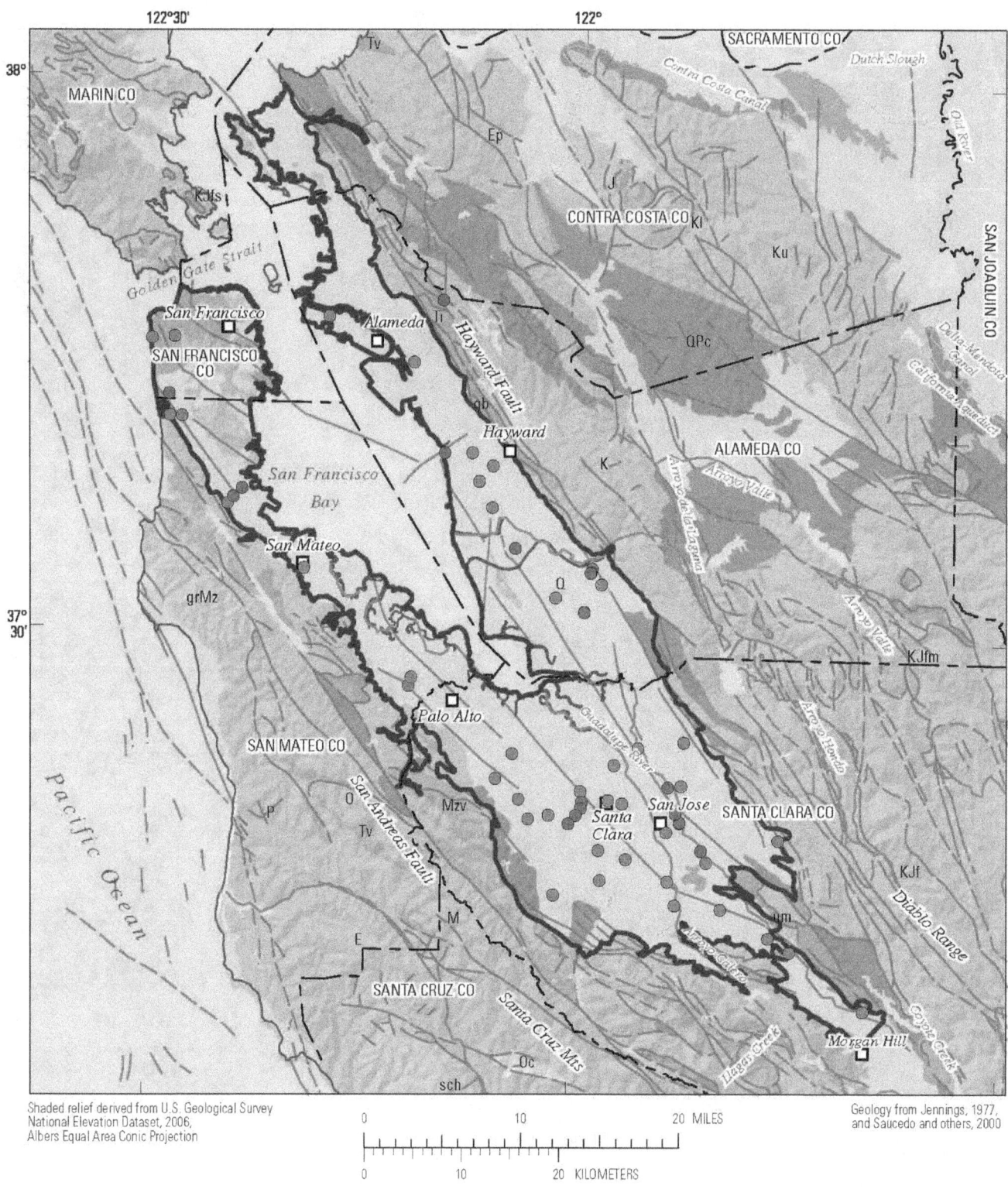

Shaded relief derived from U.S. Geological Survey
National Elevation Dataset, 2006,
Albers Equal Area Conic Projection

0   10   20 MILES

0   10   20 KILOMETERS

Geology from Jennings, 1977,
and Saucedo and others, 2000

Figure 3.   Geologic formations and areal distribution of USGS-grid and understanding wells sampled in the San Francisco Bay study unit, California GAMA Priority Basin Project.

EXPLANATION

GEOLOGIC UNIT

Cenozoic

Sedimentary rocks

Quaternary

Q	Quaternary alluvium
Qs	Quaternary other sediment
QPc	Plio-Pleistocene sediment

Tertiary

| E, P | Marine sediment |
| Oc | Nonmarine sediment |

Volcanic rocks

Tertiary

| Tv, Ti | Mafic volcanics |
| Tvp | Pyroclastic volcanics |

Mesozoic

Sedimentary and metasedimentary rocks

Cretaceous

| K, Ku | Metasediment |
| KJf(m) | Franciscan complex |

Plutonic, metavolcanic, and mixed rocks

grMz	Granitic rocks
um	Ultramafic/mafic
Mzv	Metamorphic other

| | Water |

Study unit and study area boundary

Fault—Dashed where approximately located, dotted where concealed, queried where uncertain

Aqueduct

Canal

Stream

Counties

Water boundary

USGS-grid well

USGS-understanding well

Figure 3.—Continued

The primary aquifer system that is the focus of the GAMA Priority Basin Project represents the water-bearing units that supply water for wells listed in the CDPH database. The main water-bearing units within the Visitacion Valley, Islais Valley, Westside, and South San Francisco basins consist of Pleistocene-age unconsolidated sediments of dune sand, the Pleistocene-age Colma Formation (deposits consisting of fine-grained sand, silty sand, and discontinuous beds of clay), marine estuarine deposits (locally referred to as the "Bay Mud"), and artificial fill (California Department of Water Resources, 2003). Impermeable bedrock underlying the water basins is composed of consolidated sediments of the Franciscan Group (Schlocker, 1974) and interbedded strata of marine mudstone, sandstone, and conglomerate of late Jurassic and Cretaceous age (Bailey and others, 1964; California Department of Water Resources, 2003). The water-bearing units within the Lobos, Marina, and Downtown basins include Pleistocene-age deposits consisting of fine-grained sand, silty sand, and discontinuous beds of clay, and alluvial fan deposits (California Department of Water Resources, 2003). In the Santa Clara Valley, the aquifers are composed of Holocene-age and Pleistocene-age fluvial deposits, the Bay Mud, and the Pleistocene-age Santa Clara Formation (pebbly sandstone, siltstone, and clay) where it is exposed on the west and east sides of the Santa Clara Valley (California Department of Water Resources, 2003; Hanson and others, 2004).

Sources of natural recharge to the groundwater flow system in the study unit include infiltration of mountain-front runoff, streamflow, and local precipitation. However, the predominant recharge sources for the groundwater flow system are (1) artificial recharge from the infiltration of imported water from the Hetch Hetchy Reservoir at percolation ponds, (2) leakage from transmission pipelines that transport the imported water, and (3) deep percolation of return flow from landscape irrigation (Hanson and others, 2004). Groundwater discharge occurs as pumpage, baseflow to streams, evapotranspiration, and direct discharge to the San Francisco Bay (Hanson and others, 2004). Groundwater generally flows toward the center of the basin and the San Francisco Bay.

Saltwater intrusion to the groundwater resources is a known water-quality concern in the SFBAY study unit. Due to overdraft of the groundwater resources in the Santa Clara Valley basin, land subsidence has allowed saline waters from the Bay and adjacent salt ponds to flow upstream in rivers and streams and enter the shallow aquifers (California Environmental Protection Agency, San Francisco Bay Regional Water Quality Control Board, 2003). Pumping and abandoned wells have allowed these brackish waters to infiltrate deeper into the primary aquifer system (Figuers, 1998; California Environmental Protection Agency, San Francisco Bay Regional Water Quality Control Board, 2003).

The SFBAY study unit overlies Mesozoic- and Cenozoic-age formations, resting in a synform (downward-arched fold) (California Environmental Protection Agency, San Francisco Bay Regional Water Quality Control Board, 2003). The impermeable basement beneath the water-bearing units is largely composed of the Mesozoic-age Franciscan Group, which is a large complex of highly altered submarine-deposited volcanic rocks and chert deposits. The basement also is composed of Cretaceous-age sedimentary formations, known as the Great Valley Sequence, consisting of bedded series of sandstone, siltstone, shale, and thick conglomerate layers, and the Cenozoic-age Santa Clara Formation, which is a broadly folded, lightly consolidated, fluvial unit of conglomerate, sandstone, and mudstone. The Santa Clara Formation is slightly water bearing in the areas of the study unit where the formation dips below younger alluvial deposits.

The Hayward Fault in the Diablo Range and the San Andreas Fault in the Santa Cruz Mountains form the present eastern and western boundaries of the synform, respectively. Both faults are part of a complex structural system of northwesterly trending valleys and mountain ranges, created by tectonic uplift caused by the collision of the North American Plate with the Pacific Ocean Plate. Most faults in the area are no longer active; however, the Hayward Fault and the San Andreas Fault remain prominent, active faults (Figuers, 1998). The Hayward Fault separates Mesozoic-age units (on the west) from Cenozoic-age units (on the east) and impedes the westward flow of groundwater (California Environmental Protection Agency, San Francisco Bay Regional Water Quality Control Board, 2003). The Diablo Range separates the Santa Clara Valley on the west from the San Joaquin Valley on the east. The Santa Cruz Mountains form a ridge along the San Francisco Peninsula, which separates the Pacific Ocean from the San Francisco Bay and the Santa Clara Valley. The San Andreas Fault runs near the ridgeline (average altitude of 3,000 feet) throughout the range. The Santa Cruz Mountains are the result of tectonic uplift caused by a leftward bend in the San Andreas Fault (Wallace, 1990).

The climate in the SFBAY area is Mediterranean and is characterized by warm, dry summers and cool, moist winters (Hanson and others, 2004). The average annual temperature is 61 degrees Fahrenheit (°F) in Santa Clara and 58 °F in San Francisco (U.S. Department of Commerce, 2010). Average rainfall across the study unit ranges from 14 inches per year (in/yr) in the southern and southwestern parts of the study unit to 28 in/yr in the northern parts of the study unit (PRISM Climate Group, Oregon State University, 2010); most precipitation falls between November and April. Precipitation can be as much as 90 in/yr in the surrounding mountains, falling mostly as rain with some snow in the higher elevations (California Department of Water Resources, 2003). Most of the precipitation falls in the western side of the study unit; the Santa Cruz Mountains block the delivery of moisture to the Diablo Range, which is situated approximately 40 miles inland from the ocean, and as a result receives less precipitation than the Santa Cruz Mountains.

The Coyote Creek (fig. 2) is the main drainage feature of the Santa Clara Valley groundwater basin. It originates in the Diablo Range and flows northwesterly through the valley before entering the San Francisco Bay (California Environmental Protection Agency, San Francisco Bay

Regional Water Quality Control Board, 2003). The Guadalupe River also runs through the Santa Clara Valley, originating in the Santa Cruz Mountains and entering the south end of the Bay (California Department of Water Resources, 2003). Other creeks and streams of the natural streamflow network throughout the study unit discharge directly to the Bay. The surface-water system also includes a system of reservoirs, aqueducts, and pipelines. The reservoirs discharge directly to creeks, whereas aqueducts and pipelines transfer imported water to artificial-recharge facilities (Hanson and others, 2004).

Methods

The status assessment provides a spatially unbiased assessment of groundwater quality in the primary aquifer system, while the understanding assessment was designed to evaluate the natural and human factors that may affect groundwater quality of the SFBAY study unit. This section describes the methods used to: (1) define groundwater quality, (2) assemble the datasets used for the status assessment, (3) determine which constituents warrant additional evaluation, (4) calculate aquifer-scale proportions, and (5) analyze statistics for the understanding assessment. Methods used for compilation of data on potential explanatory factors are described in appendix D.

In this study, groundwater-quality data are presented as *relative-concentrations*, the concentrations of constituents measured in groundwater relative to regulatory and non-regulatory benchmarks used to evaluate drinking-water quality. Constituents were selected for additional evaluation in the assessment on the basis of objective criteria defined in terms of relative-concentrations. Groundwater-quality data collected by USGS-GAMA and data compiled in the CDPH database are used in the status assessment. Two statistical methods based on spatially unbiased equal-area grids are used to calculate aquifer-scale proportions of low, moderate, or high relative-concentrations: (1) the "grid-based" method uses one value per grid cell to represent groundwater quality, and (2) the "spatially weighted" method uses many values per grid cell (Belitz and others, 2010).

The CDPH database contains historical records from more than 27,000 wells, necessitating targeted data retrievals to effectively access relevant water-quality data. For example, for the area representing the SFBAY study unit, the historical CDPH database contains more than 26,000 records for 400 wells (fig. 4). The CDPH data were used in three ways in the status assessment: (1) to fill in gaps in the USGS data for the grid-based calculations of aquifer-scale proportions, (2) to aid in selecting constituents for additional evaluation in the assessment, and (3) to provide the majority of the data used in the spatially weighted calculations of aquifer-scale proportions.

Relative-Concentrations and Water-Quality Benchmarks

Concentrations of constituents are presented as relative-concentrations in the status assessment:

$$\text{Relative-concentration} = \frac{\text{Sample concentration}}{\text{Benchmark concentration}}$$

Regulatory and non-regulatory benchmarks apply to treated water that is served to the consumer, not to untreated groundwater. However, to provide some context for the results, concentrations of constituents measured in the untreated groundwater were compared with benchmarks established by the U.S. Environmental Protection Agency (USEPA) and CDPH (U.S. Environmental Protection Agency, 2006; California Department of Public Health, 2008a,b). Relative-concentrations less than 1 (< 1.0) indicate sample concentrations less than the benchmark, and relative-concentrations greater than 1 (> 1.0) indicate sample concentrations greater than the benchmark. The use of relative-concentrations also permits comparison of constituents on a single scale for a wide range of concentrations. Relative-concentrations can only be computed for constituents with water-quality benchmarks; therefore, constituents lacking water-quality benchmarks are not included in the status assessment.

The benchmarks used for each constituent were selected in the following order of priority:

1. Regulatory, health-based CDPH and USEPA maximum contaminant levels (MCL-CA and MCL-US), action levels (AL-US), and treatment techniques (TT-US).

2. Non-regulatory, aesthetic-based CDPH and USEPA secondary maximum contaminant levels (SMCL-CA and SMCL-US). For constituents with both recommended and upper SMCL-CA levels, the values for the upper levels were used.

3. Non-regulatory, health-based CDPH notification levels (NL-CA), USEPA lifetime health advisory levels (HAL-US), and USEPA risk-specific doses for a risk of 1:100,000 (RSD5-US).

Note that for constituents with multiple types of benchmarks, this hierarchy may not result in selection of the benchmark with the lowest concentration. Additional information on the types of benchmarks and listings of the benchmarks for all analyzed constituents is provided by Ray and others (2009).

Toccalino and others (2004), Toccalino and Norman (2006), and Rowe and others (2007) previously used the ratio of the measured sample concentration to the benchmark concentration [either USEPA MCLs or health-based screening levels (HBSLs)], and defined this ratio as the benchmark quotient. HBSLs were not used in this report because they are not currently used as benchmarks by California drinking-water regulatory agencies. Because different water-quality

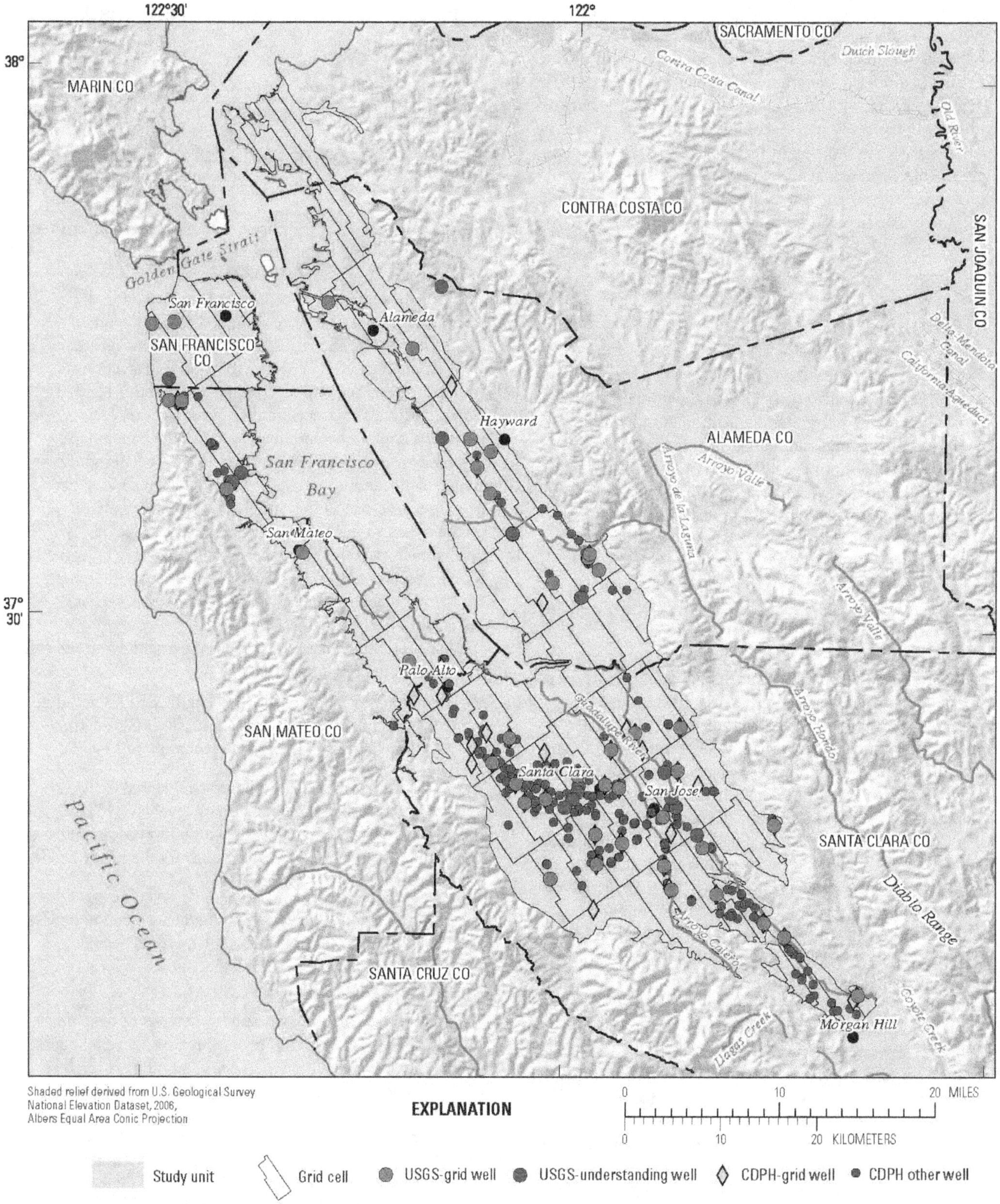

EXPLANATION

Study unit	Grid cell	● USGS-grid well	● USGS-understanding well	◇ CDPH-grid well	• CDPH other well

Figure 4. Locations of study unit grid cells, U.S. Geological Survey (USGS) grid and understanding wells, and California Department of Public Health (CDPH) wells, San Francisco Bay study unit, California GAMA Priority Basin Project, April–June 2007.

benchmarks may be used to calculate relative-concentrations and benchmark quotients, the values of these ratios may not be the same for all constituents (for example, Fram and Belitz, 2012).

For ease of discussion, relative-concentrations of constituents were classified into *low*, *moderate*, and *high* categories:

Category	Relative-concentrations for organic and special-interest constituents	Relative-concentrations for inorganic constituents
High	> 1	> 1
Moderate	> 0.1 and ≤ 1	> 0.5 and ≤ 1
Low	≤ 0.1	≤ 0.5

The boundary between "moderate" and "low" relative-concentrations was set at 0.1 for organic and special-interest constituents for consistency with other studies and reporting requirements (U.S. Environmental Protection Agency, 1998; Toccalino and others, 2004). For inorganic constituents, the boundary between "moderate" and "low" relative-concentrations was set at 0.5. The primary reason for using a higher threshold was to focus attention on the inorganic constituents of most immediate concern (Fram and Belitz, 2012). Most inorganic constituents are naturally occurring and tend to be more prevalent than organic constituents in groundwater. Although more complex classifications could be devised based upon the properties and sources of individual constituents, use of a single moderate/low boundary value for each of the two major groups of constituents provided a consistent objective criterion for distinguishing constituents occurring at moderate rather than low concentrations.

Data Used for Status Assessment

U.S. Geological Survey Grid Data

The primary data used for the grid-based calculations of aquifer-scale proportions were from wells sampled by USGS-GAMA. Detailed descriptions of the methods used to identify wells for sampling in this study unit are given in Ray and others (2009). Briefly, the study unit was divided into 10-mi^2 (~25-km^2) equal-area grid cells, and in each cell, one well was randomly selected to represent the cell (Scott, 1990). Wells were selected from the population of wells in the statewide database maintained by the CDPH. The SFBAY study unit contained 68 grid cells, and the USGS sampled wells in 43 of those cells (USGS-grid wells). Of the 43 USGS-grid wells, 30 were listed in the CDPH database; the other 13 were irrigation or domestic wells perforated at depths similar to the depths of CDPH wells in the cell. USGS-grid wells were named with an alphanumeric GAMA ID consisting of the prefix "SF" and a number indicating the relative location of the well around the San Francisco Bay (fig. A1; table A1).

Samples collected from USGS-grid wells were analyzed for 229 to 266 constituents (table 1). Water-quality indicators, VOCs, pesticides, perchlorate, nutrients, oxidation/reduction

Table 1. Number of wells sampled by the U.S. Geological Survey for the fast and slow sampling schedules, and number of constituents sampled in each constituent class for the San Francisco Bay study unit, California GAMA Priority Basin Project, April–June 2007.

[NDMA, *N*-Nitrosodimethylamine; USGS, U.S. Geological Survey; ns, not sampled]

	Schedule	
	Fast	Slow
Total number of wells	48	31
Number of grid wells sampled	40	3
Number of understanding wells sampled	8	28
Analyte groups	**Number of constituents**	
Inorganic constituents		
Specific conductance	1	1
Major and minor ions, alkalinity, and total dissolved solids	ns	11
Trace elements	ns	24
Nutrients	5	5
Organic constituents		
Volatile organic compounds (VOCs) [1]	85	85
Pesticides and degradates	63	63
Polar pesticides and metabolites [2]	54	54
Special-interest constituents		
NDMA	ns	1
Perchlorate	1	1
Geochemical and age-dating tracers		
Dissolved oxygen, temperature, pH	3	3
Arsenic, chromium, and iron redox species ratios	3	3
$\delta^2 H$ and $\delta^{18} O$ of water	2	2
$\delta^{14} N$ and $\delta^{18} O$ isotopes of nitrate	2	2
Carbon^{14} and $\delta^{13} C$ of dissolved carbonate	2	2
Noble gases (He, Ne, Ar, Kr, Xe), ^{3}He/^{4}He, and tritium [3]	7	7
Tritium [4]	1	1
Radon-222	ns	1
Sum [5]:	229	266

[1] Includes nine constituents classified as fumigants.

[2] Does not include four constituents in common with pesticides and degradates.

[3] Analyzed at Lawrence Livermore National Laboratory, Livermore, California.

[4] Analyzed at USGS Stable Isotope and Tritium Laboratory, Menlo Park, California.

[5] Fourteen pharmaceutical compounds were analyzed at slow wells, and results are discussed in Fram and Belitz, 2011.

(redox) species ratios, noble gases, and selected isotopes were analyzed in samples from all wells. Major and minor ions, trace elements, and radioactive constituents were analyzed in samples from 31 wells on the slow sampling schedule. The sampling schedules, collection, analysis, and quality-control data for the analyte groups listed in table 1 are described by Ray and others (2009). Total dissolved solids (TDS) was measured directly or calculated from specific conductance (see appendix B).

California Department of Public Health Grid Data

Data collected by USGS-GAMA at the USGS-grid wells provided part of the data used for the status assessment for inorganic constituents; the rest of the data were obtained from the CDPH database. Of the 68 grid cells in the study unit, 25 cells did not have USGS-grid-well data, and 40 cells had USGS-grid-well data but did not have USGS data for major ions, trace elements, or radioactive constituents. The CDPH database was queried to provide these missing data for inorganic constituents. CDPH wells with data for the most recent 3 years available at the time of sampling (April 1, 2004–March 31, 2007) were considered. If a well had more than one analysis for a constituent in the 3-year interval, the most recent data were selected. Selected wells from the CDPH database that were used to supplement the USGS data are referred to as CDPH-grid wells in this report.

The procedures used to identify suitable data from CDPH wells are described in appendix A. Briefly, the first choice was to use CDPH data from the same well sampled by the USGS. In this case, "DG" was added to the data's GAMA ID to signify that it was from a well sampled by the USGS and that the USGS data were supplemented from the CDPH database (fig. A1B; table A1). If the DG well did not have all of the needed data, then a second well in the cell was randomly selected from the subset of CDPH wells with data, and a new identification with "DPH" was assigned to that well (fig. A1B; table A1). The combination of the USGS-grid-well data and the CDPH-grid-well data produced a network covering 47 of the 68 grid cells in the SFBAY study unit (table A1). The remaining 21 cells had no available data (fig. 4).

The CDPH database generally did not contain data for all of the missing inorganic constituents at every CDPH-grid well; therefore, the number of wells used for the grid-based assessment was different for different inorganic constituents (table 2). Although other organizations also collect water-quality data, the CDPH database is the only statewide database of groundwater-chemistry data available for comprehensive analysis.

CDPH data were not used to supplement USGS-grid-well data for VOCs, pesticides, or special-interest constituents. A larger number of VOCs and pesticide compounds are analyzed for by the USGS-GAMA Program than are available from the CDPH database. USGS-GAMA collected samples for 85 VOCs plus 117 pesticides and pesticide degradates from all 43 grid wells in the SFBAY study unit (table 1). In addition,

laboratory reporting levels used for USGS-GAMA analyses of organic and special-interest constituents were typically one to two orders of magnitude lower than the reporting limits for analyses compiled by the CDPH (table 3).

Table 2. Inorganic constituents and associated benchmark information, and number of grid wells with USGS-GAMA data and CDPH data, for each constituent, San Francisco Bay study unit, California GAMA Priority Basin Project.

[CDPH, California Department of Public Health; N, nitrogen; SMCL, secondary maximum contaminant level; HBB, health-based benchmark (including all benchmark types except SMCL); USGS, U.S. Geological Survey]

Constituent	Number of grid wells sampled by USGS GAMA	Number of grid wells selected from CDPH
Nutrient-HBB		
Ammonia	43	0
Nitrite	43	3
Nitrate	43	4
Trace element-HBB		
Aluminum	3	29
Antimony	3	29
Arsenic	3	29
Barium	3	29
Beryllium	3	29
Boron	3	9
Cadmium	3	29
Chromium	3	28
Copper	3	29
Lead	3	25
Molybdenum	3	4
Nickel	3	29
Selenium	3	29
Strontium	3	0
Thallium	3	30
Vanadium	3	18
Minor element-HBB		
Fluoride	3	29
Trace element-SMCL		
Iron	3	29
Manganese	3	29
Silver	3	29
Zinc	3	29
Major element-SMCL		
Chloride	3	29
Sulfate	3	29
Total dissolved solids (TDS)	3	29
Radioactive-HBB		
Uranium	3	5
Radon-222	3	5

Table 3.   Comparison of the number of compounds and median laboratory reporting levels or method detection levels, by type of constituent, for data reported in the California Department of Public Health (CDPH) database and for data collected by the U.S. Geological Survey (USGS) for the San Francisco Bay study unit, California GAMA Priority Basin Project, April–June 2007.

[CDPH, California Department of Public Health; MDL, method detection level; LRL, laboratory reporting level; mg/L, milligrams per liter; μg/L, micrograms per liter; pCi/L, picocuries per liter; SSMDC, sample-specific minimum detectable concentration; nc, not collected]

Constituent type	CDPH		GAMA		Median unit
	Number of compounds	Median MDL	Number of compounds	Median LRL	
Volatile organic compounds (including fumigants)	61	0.5	85	0.06	μg/L
Pesticides plus degradates	28	2.5	117	0.019	μg/L
Nutrients, major and minor ions	12	0.4	16	0.02	mg/L
Trace elements	24	4	24	0.11	μg/L
Radioactive constituents (SSMDC) [1]	6	2	2	0.6	pCi/L
Perchlorate	1	4	1	0.5	μg/L
N-Nitrosodimethylamine (NDMA)	nc	nc	1	0.002	μg/L

[1] Value reported in GAMA column is a median SSMDC for two radioactive constituents collected and analyzed by GAMA.

Data Used for Spatially Weighted Calculation

The spatially weighted aquifer-scale proportions of relative-concentrations were calculated with data from the USGS-grid wells, from selected additional wells sampled by USGS-GAMA, and from all wells in the CDPH database having water-quality data collected during the 3-year interval April 1, 2004–March 31, 2007. For wells with both USGS and CDPH data, only the USGS data were used.

Thirty-six additional wells were sampled by the USGS to provide more data in several areas to better understand specific groundwater-quality issues in the study unit (Ray and others, 2009). These "USGS-understanding" wells included 12 production wells and 6 monitoring well clusters containing a total of 24 monitoring wells. The production wells were given GAMA IDs consisting of the prefix "SFU" and a number indicating the relative location of the well around the Bay. The monitoring wells were given GAMA IDs consisting of the prefix "SFM," a letter A–F indicating the relative location of the well around the Bay (where A is the northwest side of the Bay, and letters are assigned counterclockwise), and a number indicating relative well depth within the cluster (where 1 is the deepest well) (fig. A1A; table A1). Only the USGS-understanding production wells were used in the spatially weighted calculations. The USGS-understanding monitoring wells were not representative of the primary aquifer system, due to well completion depths and well cluster locations (i.e., in areas not near CDPH wells).

Selection of Constituents for Status and Understanding Assessments

As many as 266 constituents were analyzed in samples from SFBAY study unit wells; however, only a subset of these constituents is identified for additional evaluation in

this report. A complete list of the constituents investigated by USGS-GAMA in the SFBAY study unit can be found in Ray and others (2009). Of the 266 constituents analyzed, 137 constituents did not have benchmarks. Because relative-concentrations cannot be calculated for constituents without benchmarks, these 137 constituents were not evaluated in this report. The 129 constituents having benchmarks were assessed, and a subset of these constituents was selected for additional evaluation in the status assessment on the basis of the following criteria:

1. Constituents present at high or moderate relative-concentrations in the CDPH database within the 3-year period (April 1, 2004–March 31, 2007);

2. Constituents present at high or moderate relative-concentrations in the USGS-grid wells or USGS-understanding wells; or

3. Organic constituents having study unit detection frequencies greater than 10 percent in the USGS-grid well dataset for the study unit.

Constituents discussed in the understanding assessment had high relative-concentrations in greater than 2 percent of the primary aquifer system, or were detected in greater than 10 percent of the USGS-grid well dataset.

The CDPH database also was used to identify constituents that have been reported at high relative-concentrations historically, but not currently (table 4). The historical period was defined as from the earliest record maintained in the CDPH electronic database to March 31, 2004 (July 5, 1977, through March 31, 2004).

Constituent concentrations may be historically high, but not currently high, because of improvement of groundwater quality with time or abandonment of wells with high concentrations. Historically high concentrations

Table 4.   Historically high constituents from the California Department of Public Health (CDPH) database, July 5, 1977, to March 31, 2004, San Francisco Bay study unit, California GAMA Priority Basin Project.

[CDPH, California Department of Public Health; high, concentration greater than human-health threshold; MCL-US, U.S. Environmental Protection Agency maximum contaminant level; MCL-CA, CDPH maximum contaminant level; SMCL-CA, CDPH secondary maximum contaminant level; mg/L, milligrams per liter; μg/L, micrograms per liter]

Constituent	Threshold type	Threshold value	Units	Date of most recent high value	Number of historically high wells[1]	Number of wells with analysis[1]
Trace elements						
Antimony	MCL-US	6	μg/L	12-29-94	6	330
Cadmium	MCL-US	5	μg/L	03-01-01	7	365
Chromium	MCL-CA	50	μg/L	12-03-03	5	365
Thallium	MCL-US	2	μg/L	03-29-94	1	330
Major and minor elements						
Sulfate	SMCL-CA	500	mg/L	12-18-03	1	368
Fluoride	MCL-CA	2	mg/L	06-07-01	5	368
Solvents						
1,2-Dichloroethane	MCL-CA	0.5	μg/L	02-02-90	1	384
Fumigants						
1,2-Dichloropropane	MCL-CA	5	μg/L	12-08-97	1	380
Insecticide/degradate						
Aldicarb Sulfone	MCL-US	3	μg/L	06-16-93	1	227
Constituent of special interest						
Perchlorate	MCL-CA	6	μg/L	12-19-00	2	193

[1] Based on historical CDPH well data from July 5, 1977, to March 31, 2004.

of constituents that do not otherwise meet the criteria for inclusion in the status assessment are not considered representative of potential groundwater-quality concerns in the study unit from 2004 to 2007. For the SFBAY study unit, 10 constituents had historically high concentrations (table 4). Half of the constituents reported at high concentrations only during the historical period were reported at high concentrations in only 1 well.

Calculation of Aquifer-Scale Proportions

The status assessment is intended to characterize the quality of groundwater resources within the primary aquifer system of the SFBAY study unit. The primary aquifer system is defined by the depth intervals over which wells listed in the CDPH database are perforated; these wells primarily are classified as municipal and community public-supply wells. The use of the term "primary aquifer system" does not imply a discrete aquifer unit. In most groundwater basins, municipal and community drinking-water supply wells generally are perforated at greater depths than are domestic wells. However, to the extent that domestic wells are perforated over the same depth intervals as the CDPH wells, the assessments presented in this report also may apply to the portions of the aquifer system used for domestic drinking-water supplies.

Two statistical approaches, grid-based and spatially weighted, were applied to evaluate the proportions of the primary aquifer system in the SFBAY study unit with high, moderate, and low relative-concentrations of constituents (Belitz and others, 2010). For ease of discussion, these proportions are referred to as "high," "moderate," and "low" aquifer-scale proportions. Calculations of aquifer-scale proportions were made for individual constituents meeting the criteria for additional evaluation in the status assessment and for classes of constituents. The classes consisted of groups of related individual constituents. For constituents with human-health benchmarks, the classes included trace elements, radioactive constituents, nutrients, trihalomethanes, solvents, other VOCs, and herbicides. For constituents with aesthetic-based benchmarks, the classes included salinity indicators and trace elements.

The grid-based calculation uses the dataset assembled from the USGS-grid-well and CDPH-grid-well data. For each constituent, the high aquifer-scale proportion was calculated by dividing the number of cells that had a high relative-concentration value for that constituent by the total number of grid cells with data for that constituent (Belitz and others, 2010). The moderate and low aquifer-scale proportions were calculated similarly. Confidence intervals for the high aquifer-scale proportions were computed using the Jeffreys interval for the binomial distribution (Brown and others, 2001; Belitz and others, 2010). Although the grid-based approach is spatially unbiased, it may not detect constituents that are present at high concentrations in small proportions of the primary aquifer system. For calculation of aquifer-scale proportion for a class of constituents, cells were considered high if any of the constituents in that class had a high value. Cells were considered moderate if any of the constituents had a moderate value and if none of the values were high.

The spatially weighted calculation uses the dataset assembled from the grid wells, selected USGS-understanding wells, and all CDPH wells. For each constituent, the high aquifer-scale proportion was calculated by computing the proportion of wells with "high" values in each cell and then averaging the proportions for all of the cells (Isaaks and Srivastava, 1989; Belitz and others, 2010). The moderate aquifer-scale proportion was calculated similarly. Confidence intervals for spatially weighted detection frequencies of high concentrations are not described in this report. For calculation of the aquifer-scale proportion for a class of constituents, wells were considered high if any of the constituents had a high value. Wells were considered moderate if any of the constituents had a moderate value and if none of the constituents had a high value.

In addition, for each constituent, the raw detection frequencies of high and moderate values were calculated using the same dataset as was used for the spatially weighted calculations. However, these raw detection frequencies are spatially biased because the wells in the CDPH database are not uniformly distributed throughout the study unit (fig. 4). For example, if a constituent were present at high concentrations in a small region of the aquifer that had a high density of wells, the raw detection frequency of high values would be greater than the high aquifer-scale proportion. Raw detection frequencies are provided for reference, but were not used to assess aquifer-scale proportions (see appendix C for details of statistical methods).

The grid-based high aquifer-scale proportions were used to represent proportions in the primary aquifer system unless the spatially weighted proportions were significantly different from the grid-based values. Significantly different results were defined as follows:

- If the grid-based high aquifer-scale proportion was zero and the spatially weighted proportion was greater than zero, then the spatially weighted result was used. This situation can happen when the concentration of a constituent is high in a small fraction of the primary aquifer system.

- If the grid-based high aquifer-scale proportion was greater than zero and the spatially weighted proportion was outside the 90-percent confidence interval (based on the Jeffreys interval for the binomial distribution), then the spatially weighted proportion was used.

The grid-based moderate and low proportions were used in most cases because the reporting levels for many organic constituents and some inorganic constituents in the CDPH database were greater than the boundaries between the moderate and low categories. However, if the grid-based moderate proportion was zero and the spatially weighted proportion was greater than zero, then the spatially weighted value was used as a minimum estimate for the moderate proportion.

Understanding-Assessment Methods

The potential explanatory factors—land use, well depth, depth to the top of the well perforations, classified groundwater age, and geochemical condition (see appendix D for more details)—were analyzed in relation to constituents of interest in order to understand the physical and chemical processes occurring within the groundwater system. Statistical tests were used to identify significant correlations between the constituents of interest and potential explanatory factors. The strongest correlations for explanatory factors influencing water quality are shown graphically.

The data selected for the understanding assessment were from grid wells and selected USGS-understanding wells. Data from USGS-understanding wells that are monitoring wells were not included in the statistical tests; only data for USGS-understanding wells that are production wells were included. Statistical and graphical relations between explanatory factors and groundwater constituent concentrations were tested using either the set of grid plus understanding data or grid data only. Because the USGS-understanding wells were not randomly selected on a spatially distributed grid, these wells were excluded from analyses of relations of water quality to areally-distributed factors (land use) to avoid areal-clustering bias. Data from the USGS-understanding wells were used for analyses of relations between constituents and the vertically distributed explanatory factors (depth of well, groundwater age classification, and geochemical conditions).

Statistical Analysis

Nonparametric statistical methods were used to test hypotheses about the relation between water-quality variables and potential explanatory factors in this report. Nonparametric statistics are robust techniques that generally are not affected by outliers and do not require that the data follow any particular distribution (Helsel and Hirsch, 2002). The significance level (p) used to test hypotheses for this report was compared to a threshold value (α) of 5 percent ($\alpha = 0.05$) to evaluate whether the relation was statistically significant ($p < \alpha$). Throughout this report, the term "correlated" is used to indicate that the relation was statistically significant.

Two different statistical tests were used because the set of potential explanatory factors included both categorical and continuous variables. Groundwater age class and oxidation-reduction class were treated as categorical variables because there were a finite number of values a well could be assigned: for example, groundwater ages were classified as modern, pre-modern, or mixed. Land use, well depth, depth to top of perforations, dissolved oxygen, and pH were treated as continuous variables because there were an infinite number of values a well could be assigned: for example, land use was represented by percentages of land-use types. Concentrations of water-quality constituents were treated as continuous variables.

Relations between potential explanatory factors, and between potential explanatory factors and water-quality constituents were tested for significance. Correlations between continuous variables were evaluated using the Spearman's rho test to calculate the rank-order coefficient (ρ, rho) and the significance level of the correlation (p). Relations between categorical variables and continuous variables were evaluated using the Wilcoxon rank-sum test. The test was applied pair-wise to the groups to determine which pairs had significantly different median values of the continuous variable. For example, the relation between well depth and groundwater age was evaluated by testing for differences in median value of well depth between each of the three pairs of groundwater age classes: modern and mixed, modern and pre-modern, and mixed and pre-modern.

Potential Explanatory Factors

Brief descriptions of potential explanatory factors, including land use, physical characteristics of the wells, indicators of groundwater age, and geochemical conditions of the aquifer, are given in this section. Data sources and methodology used for assigning values for potential explanatory factors are described in appendix D.

Land Use

Land use was described by three land-use types: urban, agricultural, and natural (appendix D). Percentages of the three types were calculated for the study unit, and for areas within a radius of 500 meters (m) (500-m buffers) around wells (Johnson and Belitz, 2009). Some constituent sources are associated with land use; for example, anthropogenic constituents and some naturally occurring constituents may be introduced to groundwater at urban and (or) agricultural lands. Land use within the SFBAY study unit was 73 percent urban, 23 percent natural, and 4 percent agricultural (fig. 5A; appendix D). The urban landscape consists of residential, commercial, and industrial areas, and the large urban centers include the cities of San Francisco, San Mateo, Santa Clara,

San Jose, Hayward, and Alameda (fig. 6). Natural lands occur along the south and southwest edges of the Bay, in Santa Clara and Alameda Counties, and are mostly grasslands. Agricultural lands are primarily in the southern tip of the study unit extending north of Morgan Hill in Santa Clara County; the primary use of agricultural land is for pasture for livestock (cattle, sheep, and poultry) and hay. Average land use within the 500-m buffers around USGS wells was similar to land use in the study unit as a whole (fig. 5A). Land use around individual USGS-grid and understanding wells primarily consisted of mixtures of urban and natural land use (fig. 5B). Five USGS-grid wells were surrounded by greater than 5 percent agricultural land use (fig. 5B).

Well Depth and Depth to Top-of-Perforation

Some constituent sources are associated with depth: anthropogenic constituents are usually introduced at land surface, and some naturally occurring constituents are introduced to groundwater from deeper zones. Therefore, well depth can help determine the source of a particular constituent. Well construction information was available for 38 of the 43 grid wells sampled by the USGS in the SFBAY study unit. USGS-grid wells had depths ranging from 80 to 1,120 ft below land-surface datum (LSD); the median was 517 ft below LSD (fig. 7A; table D2). Wells deeper than 700 ft below LSD (the 25th percentile) are located in the Santa Clara Valley groundwater basin. Depths to the top-of-perforations ranged from 35 to 540 ft below LSD, with a median of 262 ft below LSD. The perforation length was as much as 831 ft, with a median of 270 ft. The USGS-understanding production wells have median well depths, depths to top-of-perforations, and perforation lengths (fig. 7B) similar to those of the grid wells.

The USGS-understanding monitor wells have a median perforation length of 20 ft, and 23 of the 24 wells have lengths less than or equal to 50 ft (fig. 7C). The median depth and depth to the top-of-perforation (329 ft below LSD and 314 ft below LSD, respectively) of the monitor wells were not significantly different than the median depth and depth to the top-of-perforation of the USGS-grid wells (Wilcoxon rank-sum test, p=0.23 and p=0.09, respectively); however, the monitor wells generally were not perforated over the same depth interval as the USGS-grid wells. Eight monitor wells were perforated at depths shallower than the 75th percentile of the depth to the top-of-perforations of the grid wells, and three monitor wells were perforated at depths deeper than the 25th percentile of the depth of the grid wells (table D2). In addition, several of the monitor well clusters were located in areas that were not near any CDPH wells (fig. A1A). These well clusters are used for various tasks, including monitoring of seawater intrusion and pumping for desalination plants. Because the monitor wells are located in depth zones and areas which are not used for public supply, they are not representative of the primary aquifer system.

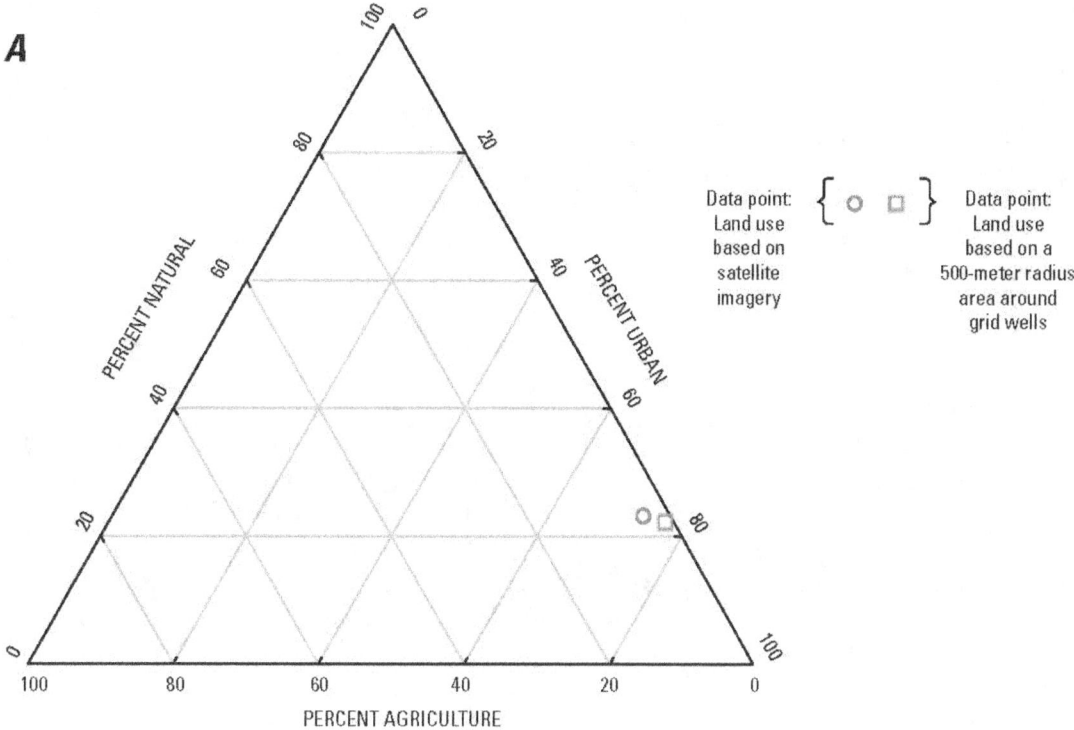

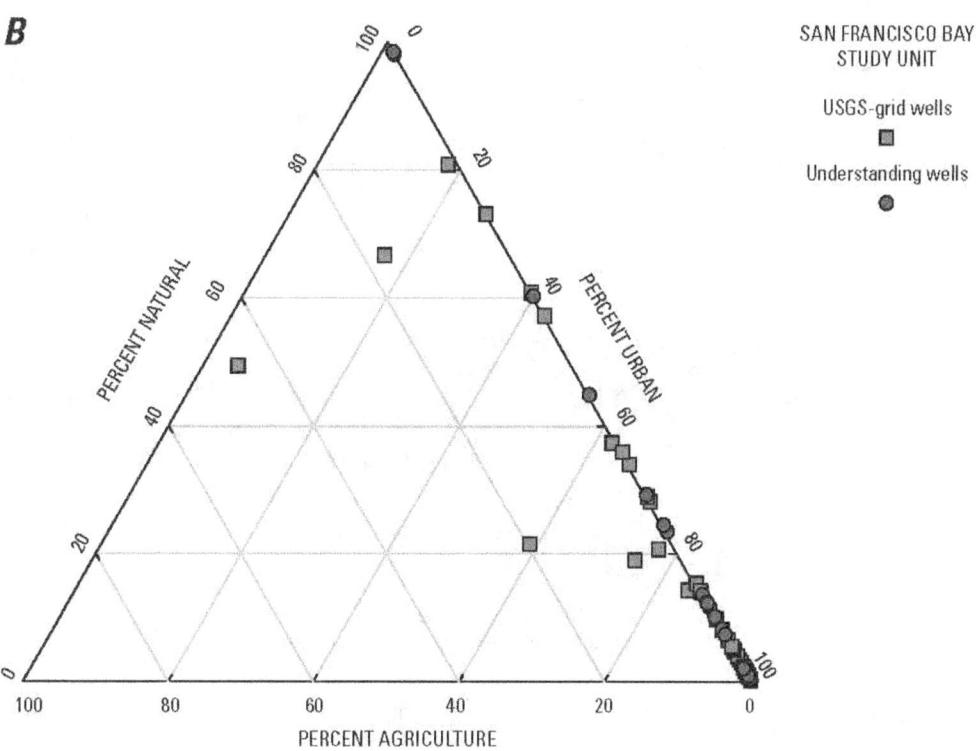

Figure 5. Percentages of urban, agricultural, and natural land use (A) in the study unit and (B) in the area surrounding each USGS-grid and USGS-understanding well in the San Francisco Bay study unit, California GAMA Priority Basin Project.

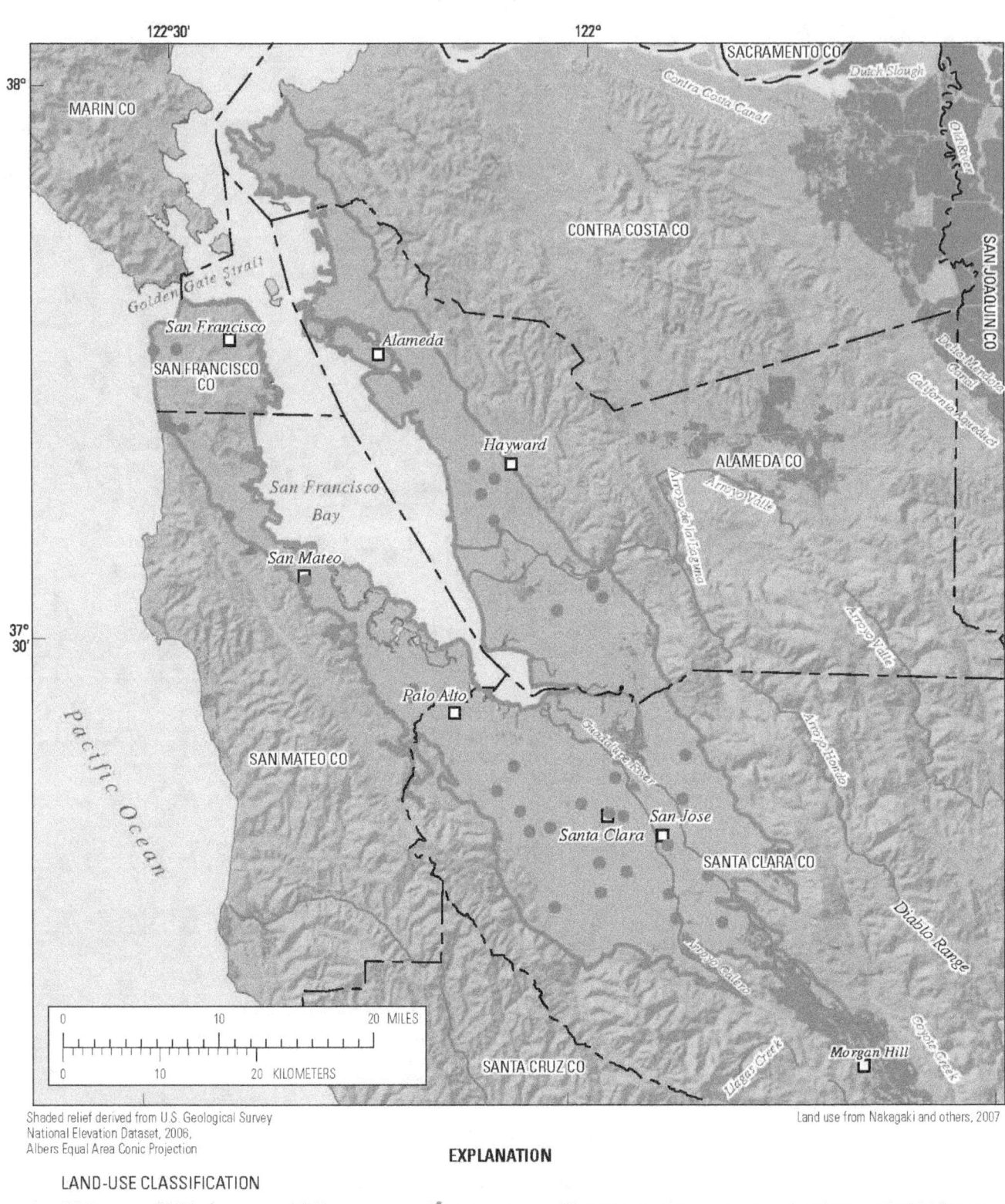

Shaded relief derived from U.S. Geological Survey
National Elevation Dataset, 2006,
Albers Equal Area Conic Projection

Land use from Nakagaki and others, 2007

EXPLANATION

LAND-USE CLASSIFICATION

Urban   Agricultural   Natural   Study unit boundary   USGS-grid well   Aqueduct   Stream

Canal   Counties

Figure 6.   Classification of land use in the San Francisco Bay study unit, California GAMA Priority Basin Project.

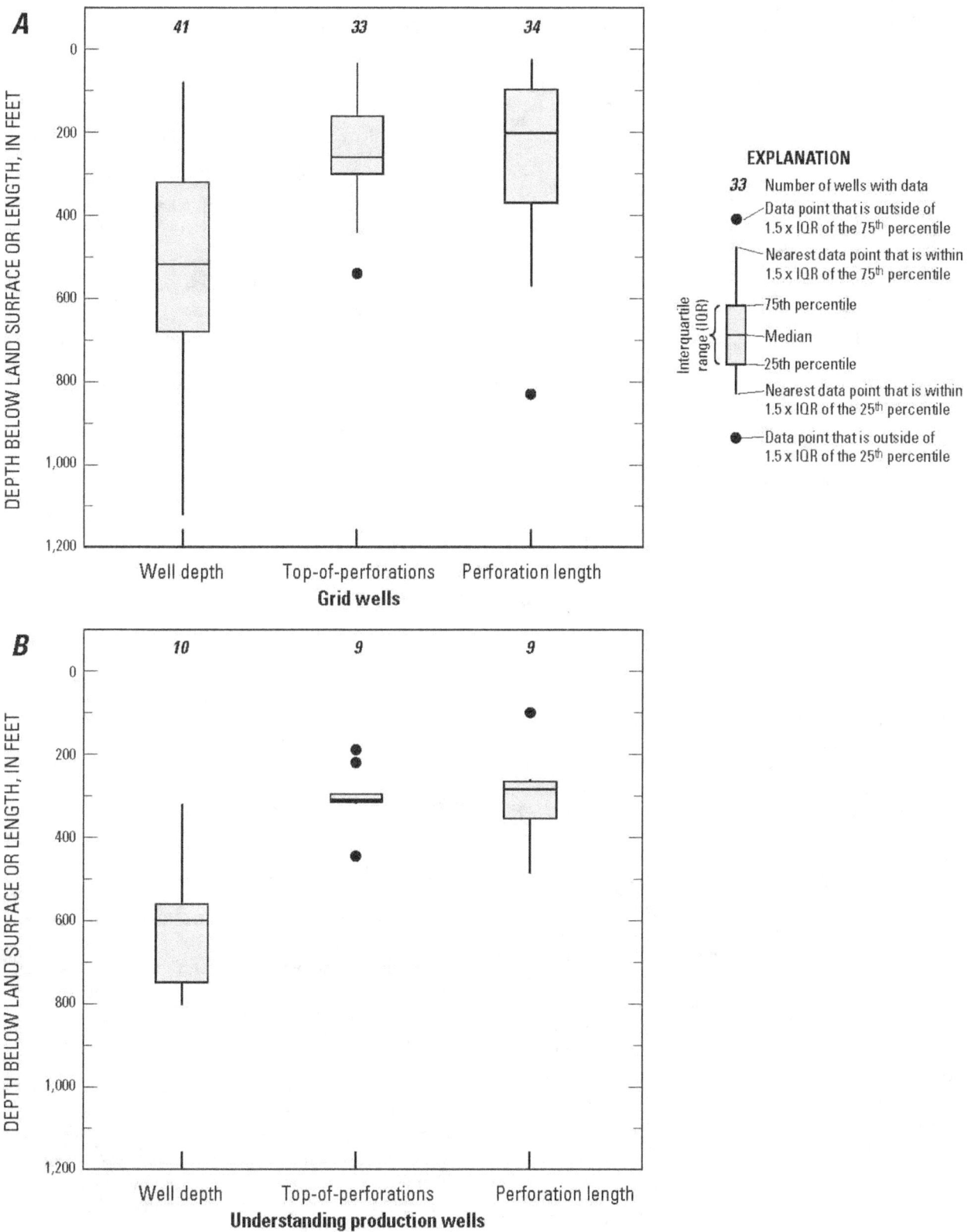

Figure 7. Well depths, depths to top-of-perforations, and perforation lengths for (*A*) grid, (*B*) understanding production, and (*C*) understanding monitor wells, San Francisco Bay Study Unit, California GAMA Priority Basin Project, April–June 2007.

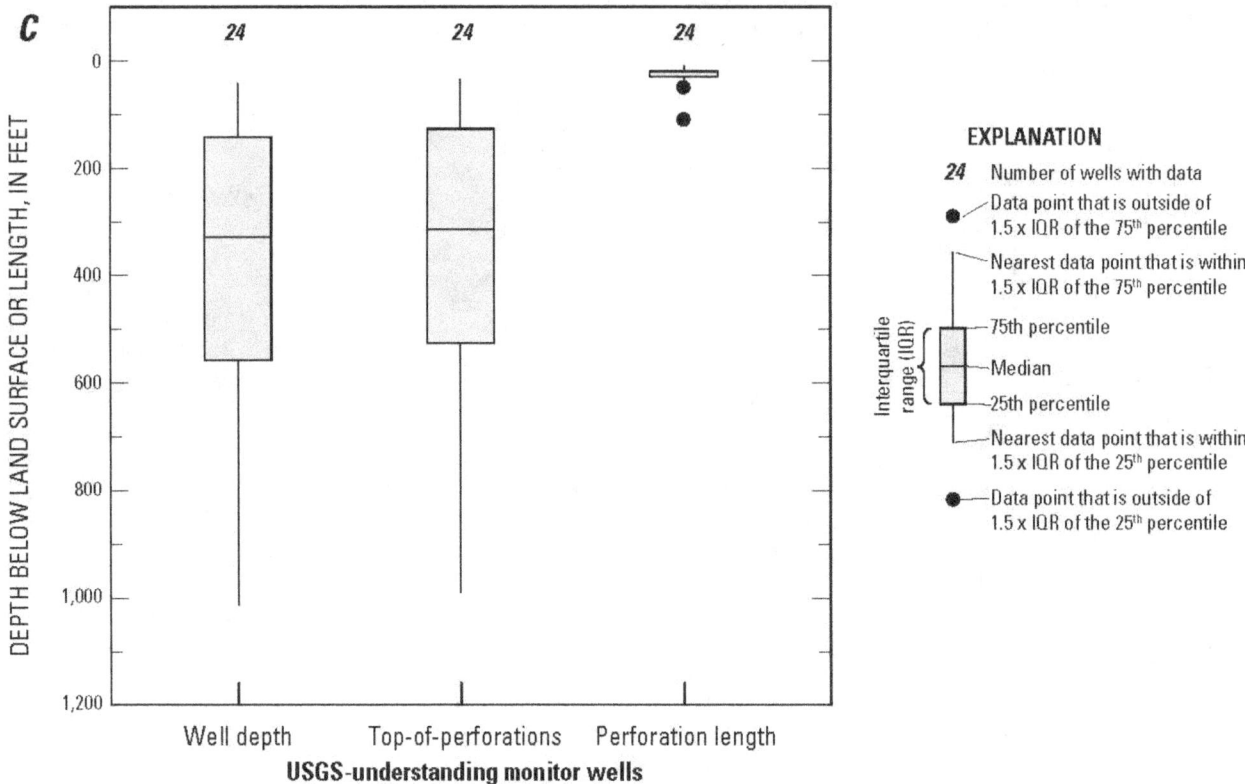

Figure 7.—Continued.

Groundwater Age

Groundwater samples were assigned age classifications based on the tritium and carbon-14 content of the samples (appendix D). Age classifications were assigned to 79 USGS-grid and understanding well groundwater samples. Among the USGS-grid and understanding production wells, 17 were classified as modern, 23 were mixed, 12 were pre-modern, and 3 were mixed or modern (table D5). In USGS-GAMA production wells, the median depths to the top-of-perforations did not increase with groundwater age (fig. 8A). Well depths or depths to top-of-perforations in USGS grid and understanding production wells were not significantly different between groundwater age classifications (fig. 8B). This suggests that wells classified as modern and pre-modern in age in the SFBAY study unit cannot be easily separated into depth classes. However, pre-modern-age groundwater was not found in USGS grid and understanding production wells perforated shallower than 300 ft below LSD (fig. 8C).

In USGS-understanding monitor wells, 2 wells were classified as modern, 9 were mixed, and 13 were pre-modern. The median well depths in monitor wells are shallower than in production wells among all age classifications. Well depths and depths to top-of-perforations in monitor wells were not significantly different between groundwater age classifications. Unlike production wells, pre-modern-age groundwater exists at all depths in monitor wells.

Geochemical Conditions

An abridged classification of oxidation-reduction (redox) conditions adapted from the framework presented by McMahon and Chapelle (2008) is given in appendix D. The classifications used in this report for 79 wells sampled by the USGS-GAMA Priority Basin Project are oxic, anoxic, or mixed (Jurgens and others, 2009) (table D3). In USGS-grid wells and understanding production wells, groundwater was 73 percent oxic, 5 percent mixed, and 22 percent anoxic. In the USGS-understanding monitor wells, groundwater was 12.5 percent oxic, 12.5 percent mixed, and 75 percent anoxic. Dissolved oxygen concentrations ranged from less than 0.2 to 11.9 milligrams per liter (mg/L). Many of the anoxic samples, including many of the monitor wells, were from the East Bay, towards the distal end of the groundwater flow path (fig. 10). Under natural conditions, the East Bay aquifers dominantly are recharged with water along the mountains, and groundwater flows westward towards the Bay. The pH ranged from 6.5 to 9.4 in the USGS-grid wells and USGS-understanding wells (fig. 9; table D3).

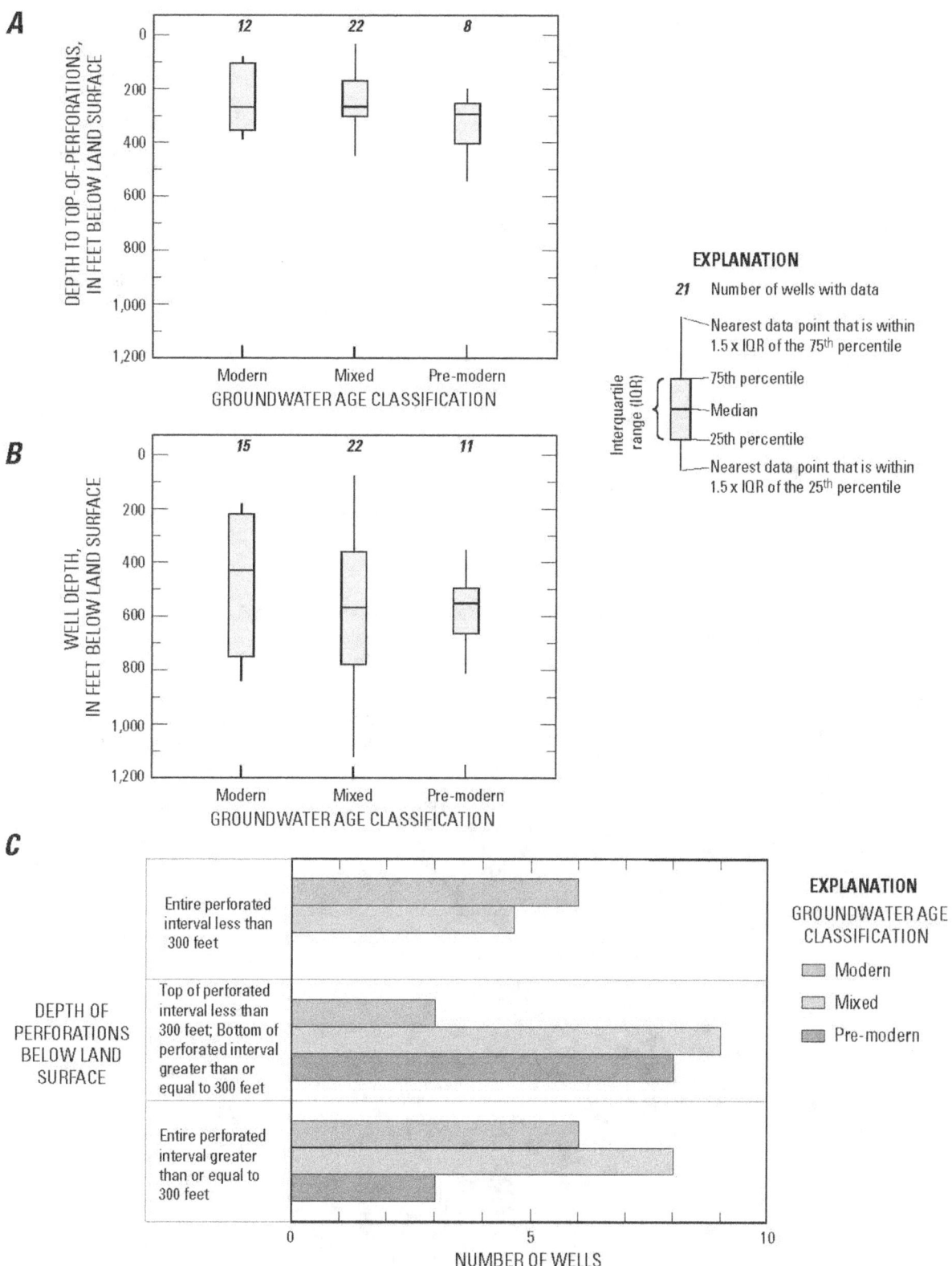

Figure 8.   Relation of groundwater age classification to (*A*) depth to top-of-perforations and to (*B*) well depth, and (*C*) bar chart showing age classification in relation to well depth classification for USGS-grid wells and USGS-understanding production wells, San Francisco Bay study unit, California GAMA Priority Basin Project, April–June 2007.

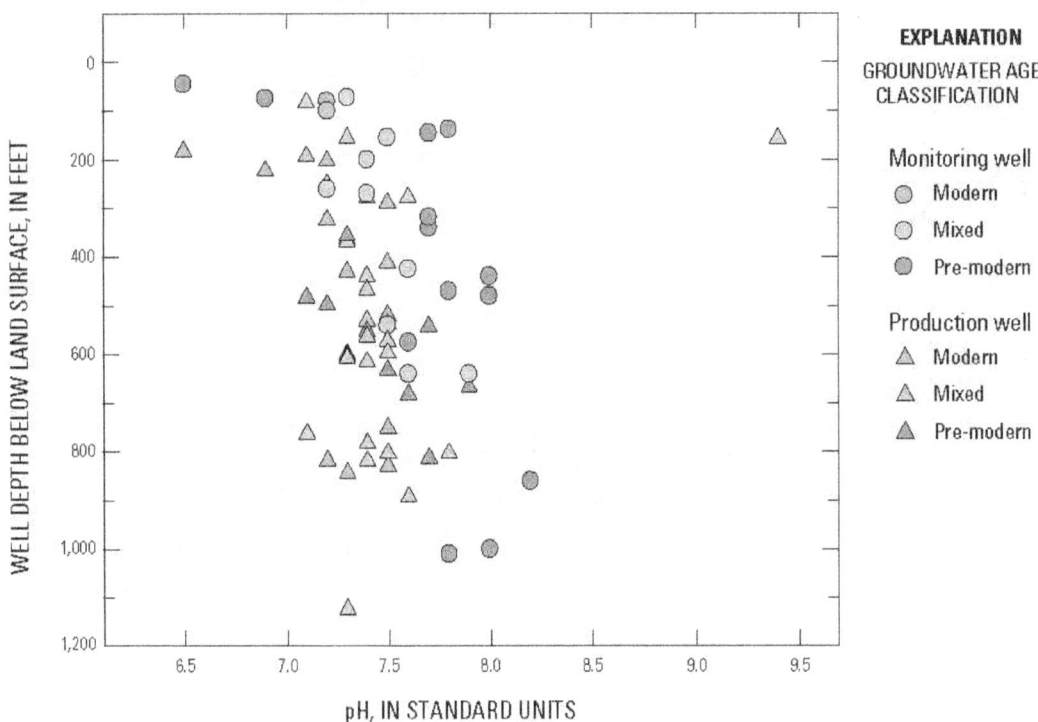

Figure 9. pH plotted as a function of well depth, well type, and age classification, San Francisco Bay Study Unit, California GAMA Priority Basin Project, April–June 2007.

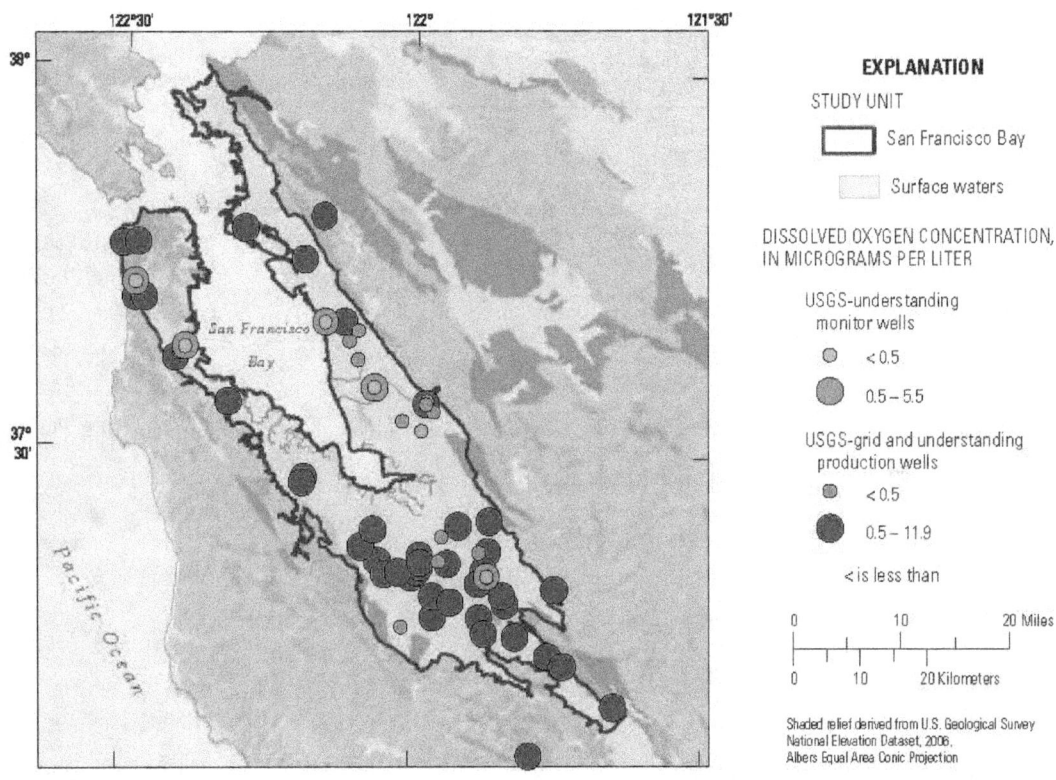

Figure 10. Dissolved oxygen concentrations in USGS-grid and USGS-understanding wells, San Francisco Bay Study Unit, California GAMA Priority Basin Project, April–June 2007.

Correlations Between Explanatory Factors

Statistical correlations between explanatory factors and water-quality constituents could reflect correlations between two or more explanatory factors presented here, among other factors. Therefore, it is important to identify statistically significant correlations between explanatory factors (table 5).

In USGS-grid wells and USGS-understanding production wells, depth to top-of-perforations is positively correlated with well depth (table 5). Depths of wells and depths to top-of-perforations were significantly deeper in wells located in urban land-use areas than in wells located in agricultural or natural land-use areas (table 5). Well depth also is positively correlated with pH, indicating greater pH in water drawn from deeper aquifers (table 5; fig. 9). Dissolved oxygen (DO) concentrations are significantly greater in mixed and modern-age groundwater than in pre-modern-age groundwater (table 6).

In USGS-understanding monitor wells, depth to top-of-perforations was greater in wells classified as mixed age than in wells classified as modern age (table 6). pH was also greater in pre-modern-age groundwater than in mixed-age groundwater. Groundwater deeper in the aquifer generally has longer residence times than groundwater from shallow sites. Implications of correlations between explanatory variables are discussed later in the report as part of the analysis of factors affecting individual constituents.

Table 5.   Results of non-parametric (Spearman's rho method) analysis of correlations in grid and understanding wells between selected potential explanatory factors, San Francisco Bay study unit, California GAMA Priority Basin Project.

[ρ, Spearman's correlation statistic; significant ($p<0.05$) positive correlation and significant negative correlations shown; nc, no significant correlation ($p>0.05$)]

Type of wells analyzed	Explanatory factors	Depth to top of perforations	Depth of well	Dissolved oxygen concentration	pH
		ρ: Spearman's correlation statistic			
Grid wells	Percent urban land use	0.48	0.63	nc	nc
Grid wells	Percent agricultural land use	−0.38	nc	nc	nc
Grid wells	Percent natural land use	−0.47	−0.63	nc	nc
Grid and understanding production wells	Depth to top of perforations				
Grid and understanding production wells	Depth of well	0.77			
Grid and understanding production wells	Dissolved oxygen concentration	nc	nc		
Grid and understanding production wells	pH	nc	0.36	nc	

Table 6. Results of Wilcoxon rank-sum tests for differences in values of selected potential explanatory factors between samples classified into groundwater age, San Francisco Bay study unit, California GAMA Priority Basin Project.

[Modern, sample with water recharged after 1952; Pre-modern, sample with water recharged before 1952; Mixed age, sample with modern and pre-modern components; only results with p-values ≤ 0.05 are considered significant in this study; Wilcoxon rank-sum tests with exact distribution and continuity correction; Z, test statistic for Wilcoxon test; significantly positive Z value (first classification is larger than second); significantly negative Z value (first classification is smaller than the second); nd, no significant difference (p-value > 0.05)]

Selected water-quality constituents	Groundwater age classification		
	Modern age compared with mixed age	Mixed age compared with pre-modern age	Modern age compared with pre-modern age
	Z: Test statistic for Wilcoxon test		
USGS-grid and understanding production wells			
Percent urban land use	nd	nd	nd
Percent agricultural land use	nd	nd	nd
Percent natural land use	nd	nd	nd
Depth to top-of-perforations	nd	nd	nd
Depth of well	nd	nd	nd
Dissolved oxygen concentration	nd	3.45	3.53
pH	nd	nd	nd
USGS-understanding monitor wells			
Percent urban land use	nd	nd	nd
Percent agricultural land use	nd	nd	nd
Percent natural land use	nd	nd	nd
Depth to top-of-perforations	−2.13	nd	nd
Depth of well	nd	nd	nd
Dissolved oxygen concentration	nd	nd	nd
pH	nd	−2.18	nd

Status and Understanding of Water Quality

The status assessment was designed to identify the constituents or classes of constituents most likely to be water-quality concerns by virtue of their high relative-concentrations or their prevalence. Approximately 24,000 individual analytical results were included in the assessment of groundwater quality for the SFBAY study unit. The status assessment applies only to constituents having regulatory or non-regulatory health-based or aesthetic/technical-based benchmarks established by the USEPA or the CDPH (as of 2009). The spatially distributed, randomized approach to well selection and data analysis yields a view of groundwater quality in which all areas of the primary aquifer system are weighted equally with respect to well density; regions with a high density of groundwater use or with a high density of potential contaminants were not preferentially sampled (Belitz and others, 2010). The understanding assessment identifies correlations between selected natural and human factors and water quality in the SFBAY study unit. The understanding assessment focuses on the constituents identified as important in the status assessment.

The following discussion of the status and understanding assessment results is divided into inorganic, organic, and special-interest constituents. The status assessment begins with a survey of how many constituents were detected at any concentration compared to the number analyzed, and a graphical summary of the relative-concentrations of constituents detected in the grid wells. Results are presented for the subset of constituents that met criteria for evaluation based on relative-concentration, or for organic constituents, prevalence.

The aquifer-scale proportions calculated using the spatially weighted approach were within the 90-percent confidence intervals for their respective grid-based aquifer high proportions for the constituents listed in table 7, indicating that the grid-based approach yields statistically equivalent results to the spatially weighted approach.

Inorganic Constituents

Inorganic constituents generally occur naturally in groundwater, although their concentrations may be influenced by human as well as natural factors. Of the 41 inorganic constituents analyzed by the USGS-GAMA Program, 38 were detected in the SFBAY study unit. Of those 38 constituents, 20 had health-based benchmarks, 6 had aesthetic-based benchmarks, and 12 had no established benchmarks (table 8). The inorganic constituents detected at high relative-concentrations in one or more grid wells were barium, iron, manganese, chloride, TDS, and nitrate. The maximum relative-concentration for each constituent is indicated in figure 11.

Table 7. Aquifer-scale proportions from grid-based and spatially weighted approaches for constituents that: (1) had moderate or high relative-concentrations during April 1, 2004, to March 31, 2007, from the California Department of Public Health (CDPH) database, (2) had moderate or high relative-concentrations in samples collected from USGS-grid wells (April–June 2007), or (3) are organic constituents having detection frequencies greater than 10 percent in USGS-grid wells, San Francisco Bay study unit, California GAMA Priority Basin Project.

[High, concentrations greater than benchmark; moderate, concentrations less than benchmark and greater than or equal to 0.1 (for organic constituents) or 0.5 (for inorganic constituents) of benchmark; low, concentrations less than 0.1 (for organic constituents) or 0.5 (for inorganic constituents) of benchmark; MCL-US, U.S. Environmental Protection Agency maximum contaminant level; MCL-CA, CDPH maximum contaminant level; NL-CA, CDPH notification level; AL-US, U.S. Environmental Protection Agency (USEPA) action level; HAL-US, USEPA Lifetime Health Advisory; SMCL-CA, CDPH secondary maximum contaminant level; mg/L, milligrams per liter; µg/L, micrograms per liter; na, not available]

Constituent	Threshold type	Threshold value	Units	Raw detection frequency [1]			Spatially weighted aquifer proportions [1]			Grid-based aquifer proportions			90-percent confidence interval for grid-based high proportion (percent) [2]	
				Number of wells	Percent moderate	Percent high	Number of cells	Proportion moderate (percent)	Proportion high (percent)	Number of cells	Proportion moderate (percent)	Proportion high (percent)	Lower limit	Upper limit
Trace elements														
Aluminum	MCL-CA	1,000	µg/L	255	0.4	0.4	34	0.2	2.9	33	0	0	0	4.0
Arsenic	MCL-US	10	µg/L	255	0.8	0	34	2.1	0	33	3.0	0	0	4.0
Barium	MCL-CA	1,000	µg/L	255	0	0.4	34	0	1.5	33	0	3.0	0.5	11
Boron	NL-CA	1,000	µg/L	50	4.0	0	14	3.6	0	12	0	0	0	11
Copper	AL-US	1,300	µg/L	253	0	0.4	35	0	0.4	33	0	0	0	4.0
Lead	AL-US	15	µg/L	215	0	0.9	32	0	1.6	29	0	0	0	4.5
Thallium	MCL-US	2	µg/L	255	0.4	0	34	0.6	0	33	0	0	0	4.0
Radioactive constituents														
Radium activity	MCL-US	5	pCi/L	109	1.8	0	22	0.9	0	na[3]	na	na	na	na
Nutrients														
Nitrate	MCL-US	10	mg/L	281	18	2.8	47	20	3.7	47	19	2.1	0.4	8.0
Nitrite	MCL-US	1	mg/L	274	0	0.4	46	0	0.3	46	0	0	0	2.9
Inorganic constituents with aesthetic benchmarks														
Chloride	SMCL-CA	500	mg/L	252	2.4	1.2	34	3.5	3.7	33	0	6.1	1.8	16
Total dissolved solids (TDS)	SMCL-CA	1,000	mg/L	297	21	1.7	47	28	5.3	43	37	7.0	1.8	16
Iron	SMCL-CA	300	µg/L	254	6.7	5.5	34	6.1	9.7	33	6.1	3.0	0.5	11
Manganese	SMCL-CA	50	µg/L	254	3.5	10	34	5.5	20	33	6.1	12	5.2	24
Trihalomethanes (THM)														
Chloroform (trichloromethane)	MCL-US	80	µg/L	263	0.4	0	46	2.2	0	43	2.3	0	0	3.1
Solvents														
1,1,1-Trichloroethane (TCA)	MCL-CA	200	µg/L	265	0	0	46	0	0	43	0	0	0	3.1

Table 7. Aquifer-scale proportions from grid-based and spatially weighted approaches for constituents that: (1) had moderate or high relative-concentrations during April 1, 2004, to March 31, 2007, from the California Department of Public Health (CDPH) database, (2) had moderate or high relative-concentrations in samples collected from USGS-grid wells (April–June 2007), or (3) are organic constituents having detection frequencies greater than 10 percent in USGS-grid wells, San Francisco Bay study unit, California GAMA Priority Basin Project.—Continued

[High, concentrations greater than benchmark; moderate, concentrations less than benchmark and greater than or equal to 0.1 (for organic constituents) or 0.5 (for inorganic constituents) of benchmark; low, concentrations less than 0.1 (for organic constituents) or 0.5 (for inorganic constituents) of benchmark; MCL-US, U.S. Environmental Protection Agency maximum contaminant level; MCL-CA, CDPH maximum contaminant level; NL-CA, CDPH notification level; AL-US, U.S. Environmental Protection Agency (USEPA) action level; HAL-US, USEPA Lifetime Health Advisory; SMCL-CA, CDPH secondary maximum contaminant level; mg/L, milligrams per liter; µg/L, micrograms per liter; na, not available]

Constituent	Threshold type	Threshold value	Units	Raw detection frequency[1]			Spatially weighted aquifer proportions[1]			Grid-based aquifer proportions			90-percent confidence interval for grid-based high proportion (percent)[2]	
				Number of wells	Percent moderate	Percent high	Number of cells	Proportion moderate (percent)	Proportion high (percent)	Number of cells	Proportion moderate (percent)	Proportion high (percent)	Lower limit	Upper limit
Solvents—Continued														
Tetrachloroethene (PCE)	MCL-US	5	µg/L	265	0.4	0.4	46	2.2	0.3	43	2.3	0	0	3.1
Trichloroethene (TCE)	MCL-US	5	µg/L	265	0.8	0	46	2.4	0	43	2.3	0	0	3.1
Other VOCs														
1,1,2-Trichlorotrifluoroethane (CFC-113)	MCL-CA	1,200	µg/L	263	0	0	46	0	0	43	0	0	0	3.1
1,1-Dichloroethene	MCL-CA	6	µg/L	265	0.4	0	46	0.4	0	43	2.3	0	0	3.1
Methyl *tert*-butyl ether (MTBE)	MCL-CA	13	µg/L	267	0	0.4	46	0	0.3	43	0	0	0	3.1
Herbicides														
Atrazine	MCL-CA	1	µg/L	190	0.4	0	46	0	0	43	2.3	0	0	3.1
Constituents of special interest														
Perchlorate	MCL-CA	6	µg/L	79	32	0	45	28	0	43	42	0	0	3.1

[1] Based on the most recent data for each CDPH well during the period April 1, 2004–March 31, 2007, combined with GAMA grid and understanding production well data.

[2] Based on the Jeffreys interval for the binomial distribution (Brown and others, 2001).

[3] USGS GAMA did not collect samples for analysis of radium activity in the SFBAY study unit.

Table 8. Numbers of constituents analyzed and detected by the U.S. Geological Survey, with associated benchmarks in each constituent class, San Francisco Bay study unit, California GAMA Priority Basin Project, April–June 2007.

[VOCs, volatile organic compounds; NWQL, National Water Quality Laboratory; USEPA, U.S. Environmental Protection Agency; CDPH, California Department of Public Health; MCL, USEPA or CDPH maximum contaminant level; HAL, USEPA health advisory level; NL, CDPH notification level; RSD5, USEPA risk specific dose at 10^{-5}; AL, USEPA action level; SMCL, USEPA or CDPH secondary maximum contaminant level]

Groups of organic constituents

Benchmark type	Sum organic and special interest		VOCs		Pesticides and degradates, and polar pesticides and metabolites		Special interest	
	Number of constituents		Number of constituents		Number of constituents		Number of constituents	
	Analyzed	Detected at any concentration	Analyzed	Detected at any concentration	Analyzed	Detected at any concentration	Analyzed	Detected at any concentration
MCL	46	18	33	14	12	3	1	1
HAL	31	4	7	2	24	2	0	0
NL	15	3	14	2	0	0	1	1
RSD5	8	0	4	0	4	0	0	0
AL	0	0	0	0	0	0	0	0
SMCL	0	0	0	0	0	0	0	0
None	104	7	27	4	77	3	0	0
Total:	204	32	85	22	117	8	2	2

Groups of inorganic constituents

Benchmark type	Sum of inorganic		Major and minor ions		Nutrients		Trace elements		Radioactive	
	Number of constituents		Number of constituents		Number of constituents		Number of constituents		Number of constituents	
	Analyzed	Detected at any concentration	Analyzed	Detected at any concentration	Analyzed	Detected at any concentration	Analyzed	Detected at any concentration	Analyzed	Detected at any concentration
MCL	15	13	1	1	2	2	11	9	1	1
HAL	3	3	0	0	1	1	2	2	0	0
NL	2	2	0	0	0	0	2	2	0	0
RSD5	0	0	0	0	0	0	0	0	0	0
AL	2	2	0	0	0	0	2	2	0	0
SMCL	7	6	3	3	0	0	4	3	0	0
None	12	12	7	7	2	2	3	3	0	0
Total:	41	38	11	11	5	5	24	21	1	1

Organic, special interest, inorganic total[1]: 245 70

[1] Does not include geochemical and age-dating tracers listed in table 1.

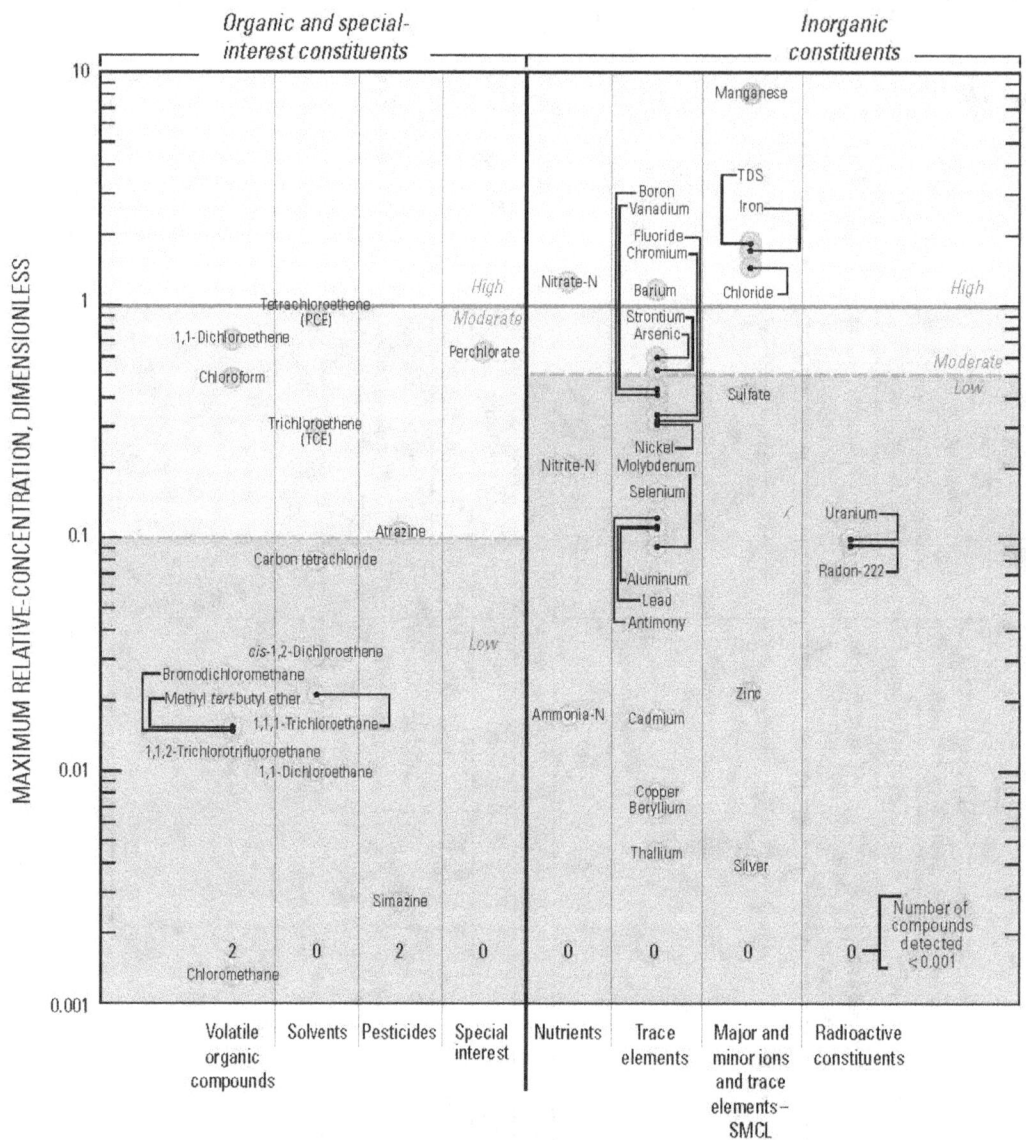

Figure 11. Maximum relative-concentration of constituents detected in grid wells, by constituent class, San Francisco Bay study unit, California GAMA Priority Basin Project, April–June 2007.

Fourteen inorganic constituents—aluminum, arsenic, barium, boron, copper, lead, thallium, nitrate, nitrite, chloride, TDS, iron, manganese, and radium—met the selection criterion of having maximum relative-concentrations > 0.5 (moderate or high) in the grid-based aquifer proportions or in the CDPH database for the period April 1, 2004, to March 31, 2007 (table 7). The percentages of the primary aquifer system with high and moderate relative-concentrations for the individual inorganic constituents are listed in table 7. Inorganic constituents, as a group (trace elements, nutrients, and radioactive constituents), had high proportions in 5.1 percent of the primary aquifer system, moderate proportions in 25 percent of the primary aquifer system, and low proportions in 70 percent of the primary aquifer system (table 9).

Inorganic constituents with relative-concentrations greater than 0.5 in one or more of the grid wells are shown in figure 12. The spatial distributions of selected inorganic constituents are illustrated with maps for USGS-grid wells, USGS-understanding wells, and CDPH-grid wells for the period April 1, 2004, to March 31, 2007 (fig. 13).

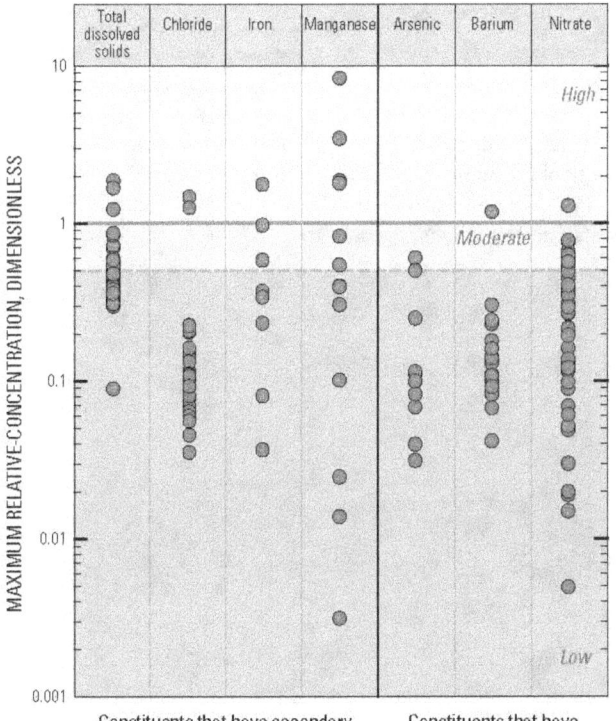

Figure 12. Maximum relative-concentrations of selected inorganic constituents in grid wells, San Francisco Bay Study Unit, California GAMA Priority Basin Project, April–June 2007.

Table 9A. Aquifer-scale proportions for constituent classes of inorganics with health-based and aesthetic benchmarks, San Francisco Bay study unit, California GAMA Priority Basin Project.

[SMCL, secondary maximum contaminant level; values are grid based except where footnoted]

Constituent class	Aquifer-scale proportion		
	Low relative-concentration (percent)	Moderate relative-concentration (percent)	High relative-concentration (percent)
Inorganics with health-based benchmarks			
Trace elements	94	3.0	3.0
Radioactive	99	[1]0.9	0
Nutrients	79	19	2.1
Any inorganic with health-based benchmarks	70	25	5.1
Inorganics with aesthetic benchmarks			
Salinity indicators	56	37	7.0
Manganese and (or) iron (SMCL)	76	12	12
Any inorganic with aesthetic benchmark	53	33	14

[1] Spatially weighted value

Table 9B. Aquifer-scale proportions for constituent classes of organics with health-based benchmarks, San Francisco Bay study unit, California GAMA Priority Basin Project.

[VOCs, volatile organic compounds; values are grid based except where footnoted]

Constituent class	Low relative-concentration (percent)		Moderate relative-concentration (percent)	High relative-concentration (percent)
	Not detected	Detected low		
Organics with health-based benchmarks				
Trihalomethanes	70	28	2.3	0
Solvents	72	23	4.7	[1]0.3
Other VOCs	81	17	2.3	[1]0.3
Herbicides	86	12	2.3	0
Any organic with health-based benchmarks	49	39	12	[1]0.6

[1] Spatially weighted value

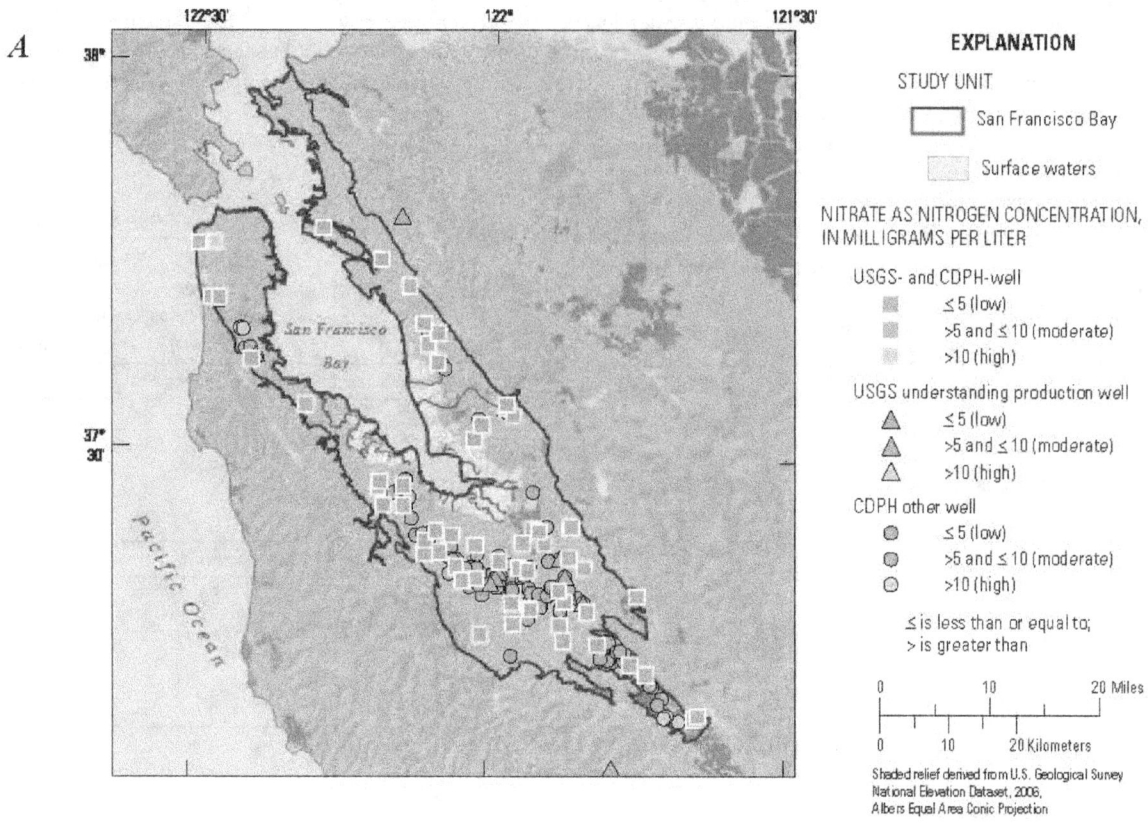

Figure 13. Concentrations of selected inorganic constituents for USGS-grid and USGS-understanding wells, April–June 2007, and from the California Department of Public Health (CDPH) database for the period April 1, 2004–March 31, 2007, San Francisco Bay study unit, California GAMA Priority Basin Project: (*A*) nitrate, (*B*) total dissolved solids, (*C*) chloride, (*D*) iron, and (*E*) manganese.

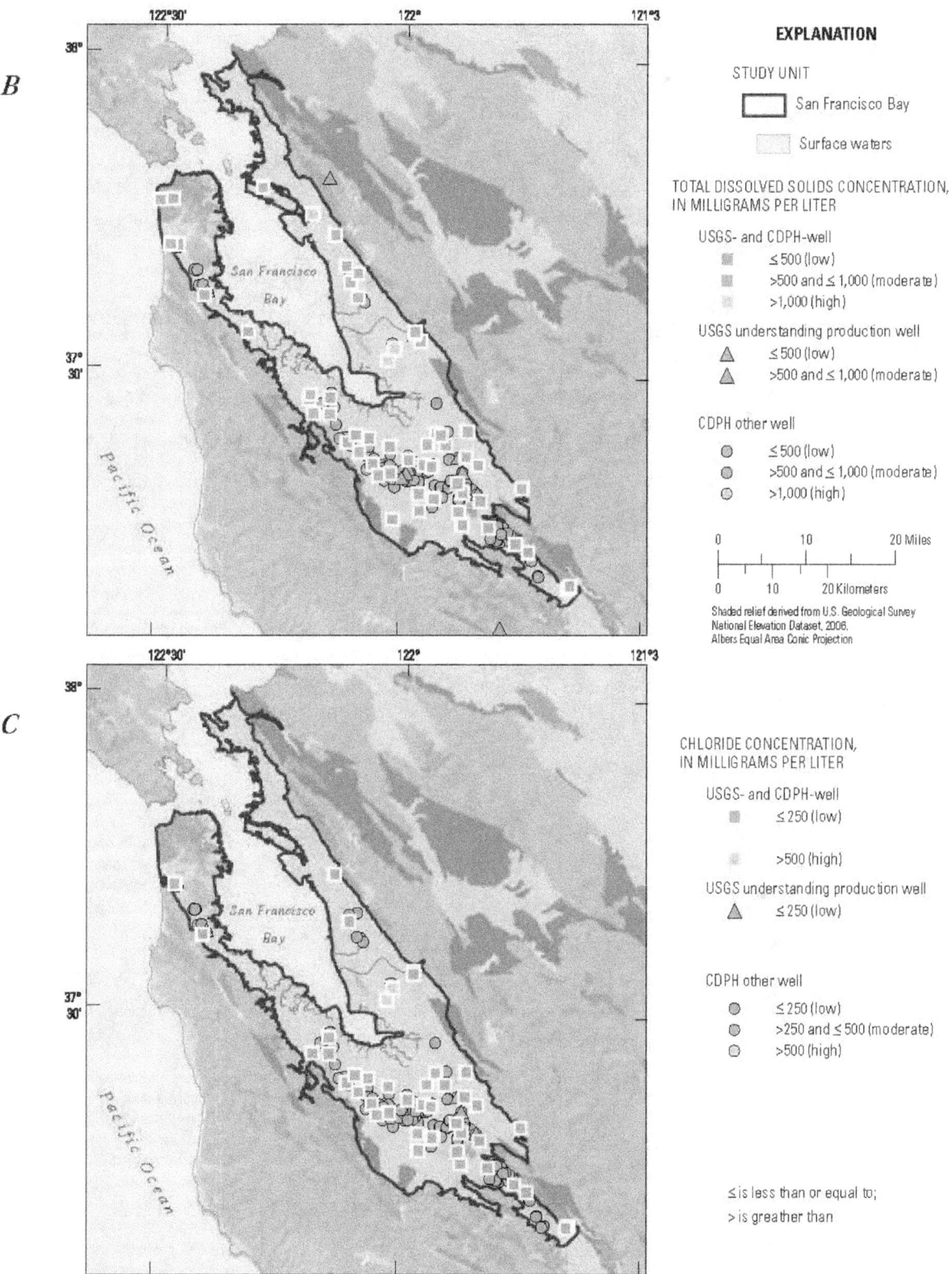

Figure 13.—Continued.

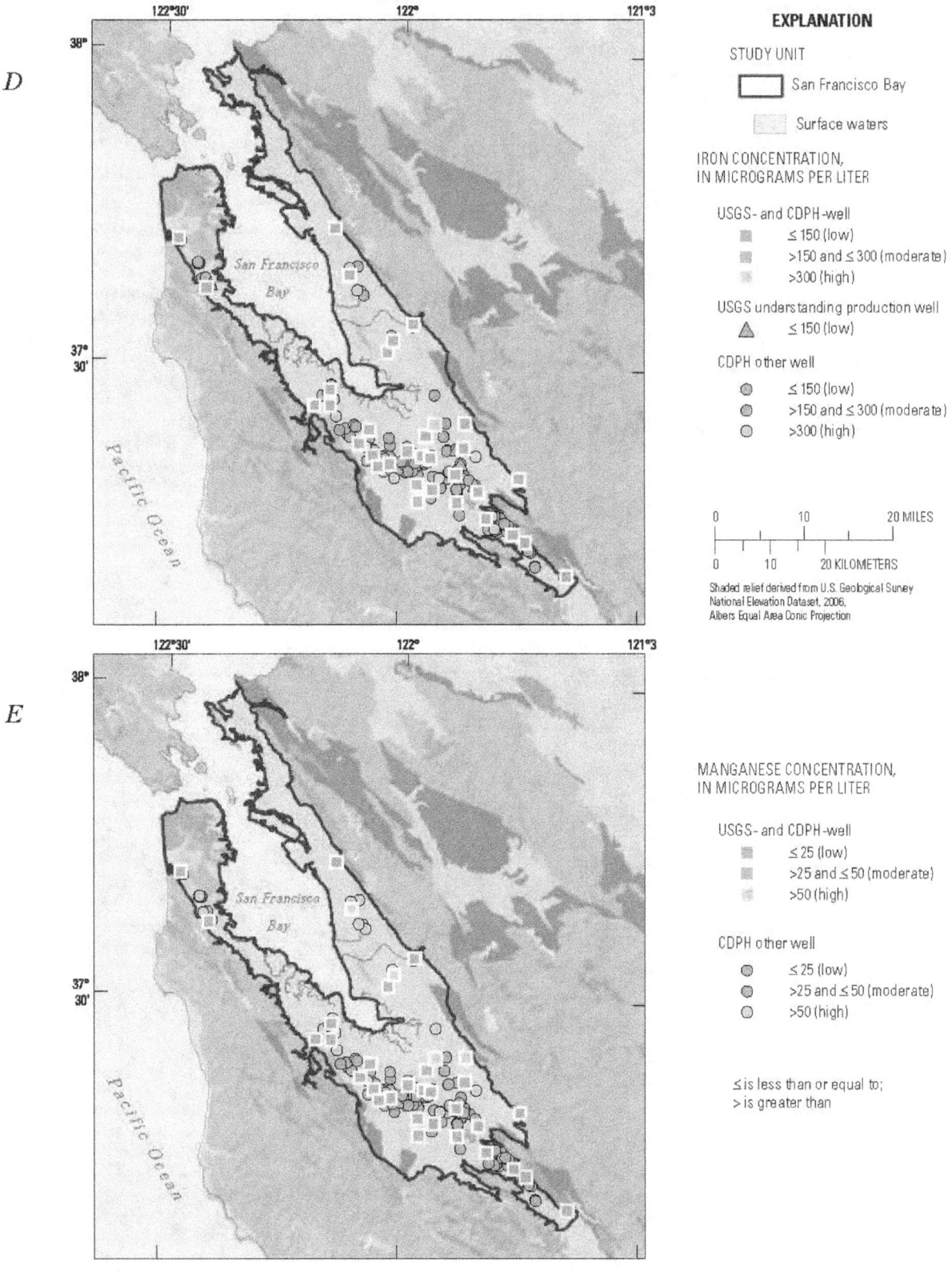

Figure 13.—Continued.

Trace Elements

Trace elements, as a class, had high relative-concentrations (for one or more constituents) in 3.0 percent of the primary aquifer system, moderate values in 3.0 percent, and low values in 94 percent (table 9). High relative-concentrations of trace elements reflected a high relative-concentration of barium (tables 7, 9) in one grid well. Moderate relative-concentrations of trace elements reflected moderate relative-concentrations of arsenic. Because there was only one well with high relative-concentration of barium in the entire dataset, no understanding assessment will be made.

Three trace elements—aluminum, copper, and lead—had spatially weighted high relative-concentrations in 2.9 percent, 0.4 percent, and 1.6 percent, respectively, compared to 0 percent for these elements for the grid-based approach (table 7). The spatially weighted approach includes data from a larger number of wells than were used in the grid-based approach, and therefore is more likely to include constituents that exist in very small proportions of an aquifer. Strontium was sampled for in only 8 of the 47 cells in the grid network, and no data were available from the CDPH database; thus, strontium was not evaluated in the status assessment.

The trace elements antimony, cadmium, chromium, and thallium had high relative-concentrations in at least one well reported in the CDPH database before April 2004 but not during the current period of study (table 4). Because these high relative-concentrations represented historical values rather than current values, these trace elements are not of concern for present-day conditions and were not selected for additional evaluation in the status assessment.

Uranium and Radioactive Constituents

In addition to uranium, the USGS sampled one radioactive constituent—radon-222—for the status and understanding assessments. The relative-concentrations of these radioactive constituents were low in the SFBAY study unit (table 9). No detections of uranium or radon-222 were above their respective MCL-US values of 30 µg/L and 4,000 picocuries per liter (pCi/L) during the current period of study or in the historical data (table 4). Radium had a moderate spatially weighted relative-concentration of 0.9 percent, but was not sampled for by the USGS in SFBAY wells.

Nutrients

Nutrients as a class had high relative-concentrations in 2.1 percent of the primary aquifer system and moderate relative-concentrations in 19 percent (table 9). The nutrient detected at high relative-concentrations was nitrate (table 7; fig. 11). High relative-concentrations of nitrate were detected on the west side of the San Francisco Peninsula and at the southern tip of the Santa Clara valley (fig. 13A).

Understanding Assessment for Nitrate

Nitrate concentrations were significantly greater in wells having water of modern or mixed ages compared with pre-modern age (table 10; fig. 14A). Nitrate was positively correlated with DO, and nitrate concentrations were significantly greater in wells with oxic conditions compared with anoxic conditions and mixed redox conditions compared with anoxic conditions (table 10; fig. 14B). Nitrate was positively correlated with percentage of urban land use and negatively correlated with percentage of natural land use (table 11; fig.14C). The negative correlation between nitrate and natural land use (table 11) suggests that the nitrate is likely from anthropogenic sources.

Nitrate was not correlated with well depth or depth to top-of-perforations (table 11, fig. 14C), and there were no significant differences in well depth or depth to top-of-perforations between wells with high or moderate relative-concentrations of nitrate and wells with low relative-concentrations of nitrate. Nitrate concentrations were significantly higher in modern and mixed-age groundwater compared to pre-modern-age groundwater, but because age classifications in the SFBAY study unit do not yield clear depth classes, moderate and high relative-concentrations of nitrate occur at varying depths of the aquifer.

Nitrate has both natural and anthropogenic sources to groundwater; however, concentrations greater than 2 mg/L (relative-concentration of 0.2) generally are considered to indicate presence of nitrate from anthropogenic sources (Mueller and Helsel, 1996). Potential anthropogenic sources of nitrate include use of fertilizers in agricultural and urban areas, nitrate in water used for engineered recharge, seepage from septic and sewage systems, and animal and human wastes. Most of the wells with high or moderate relative-concentrations of nitrate were located in the southern half of the study unit, in the Santa Clara Valley (fig. 13A). Although most of the Santa Clara Valley is currently dominated by urban land use, the valley was dominated by agricultural land use prior to World War II, and there were still extensive tracts of orchards remaining in the 1980s. Distinguishing between urban and agricultural sources of nitrate is beyond the scope of this report.

Four USGS-understanding monitor wells (three well clusters) had moderate or high relative-concentrations of nitrate. All four wells had a mixed-age classification and were in areas with production wells with moderate or high relative-concentrations of nitrate.

A

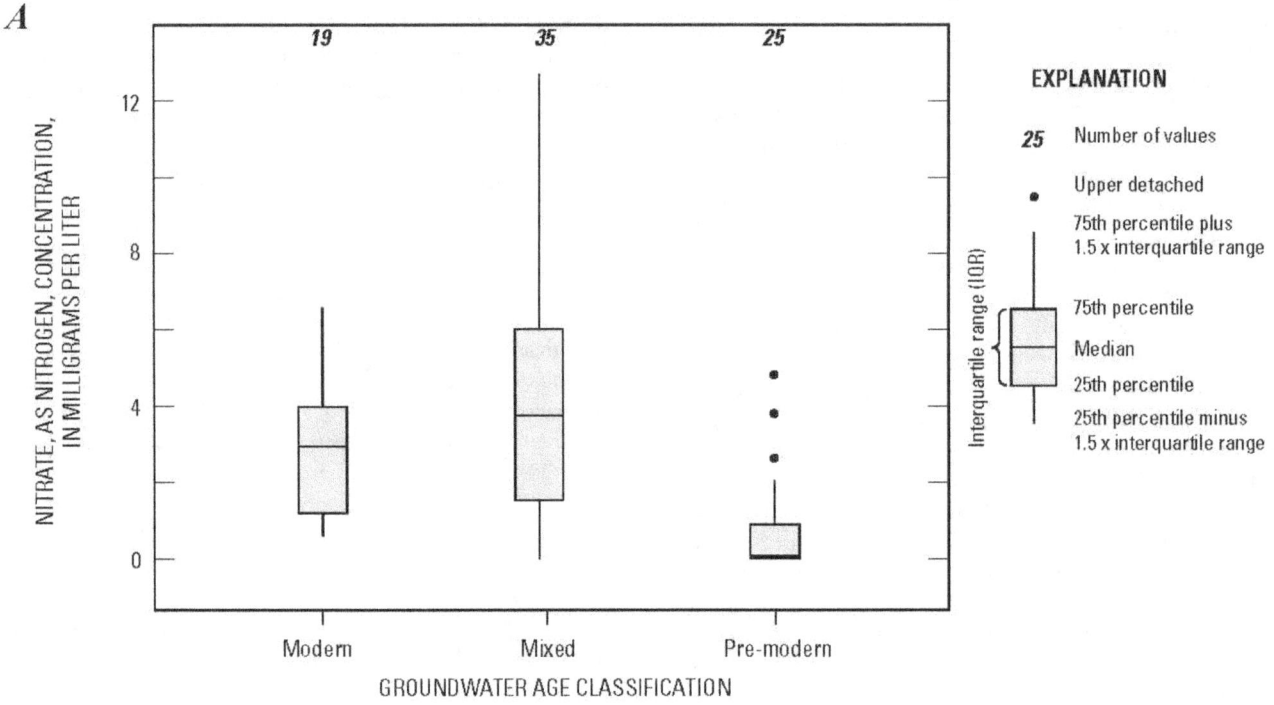

B

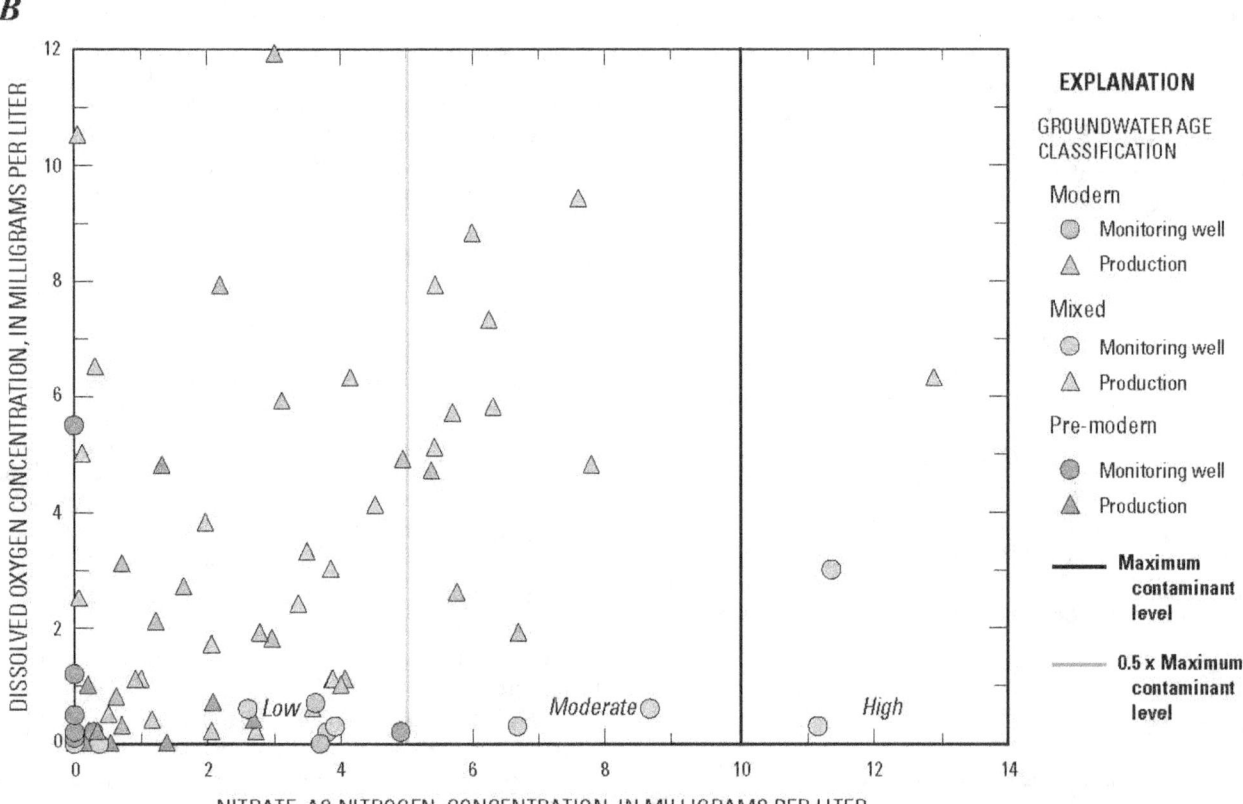

Figure 14. Nitrate, as nitrogen, concentrations related to: (*A*) classifications of groundwater age, (*B*) dissolved oxygen, well type, and classification of groundwater age, and (*C*) depth to top-of-perforations, well type, and land use in USGS-grid and USGS-understanding wells sampled for the San Francisco Bay study unit, California GAMA Priority Basin Project, April–June 2007.

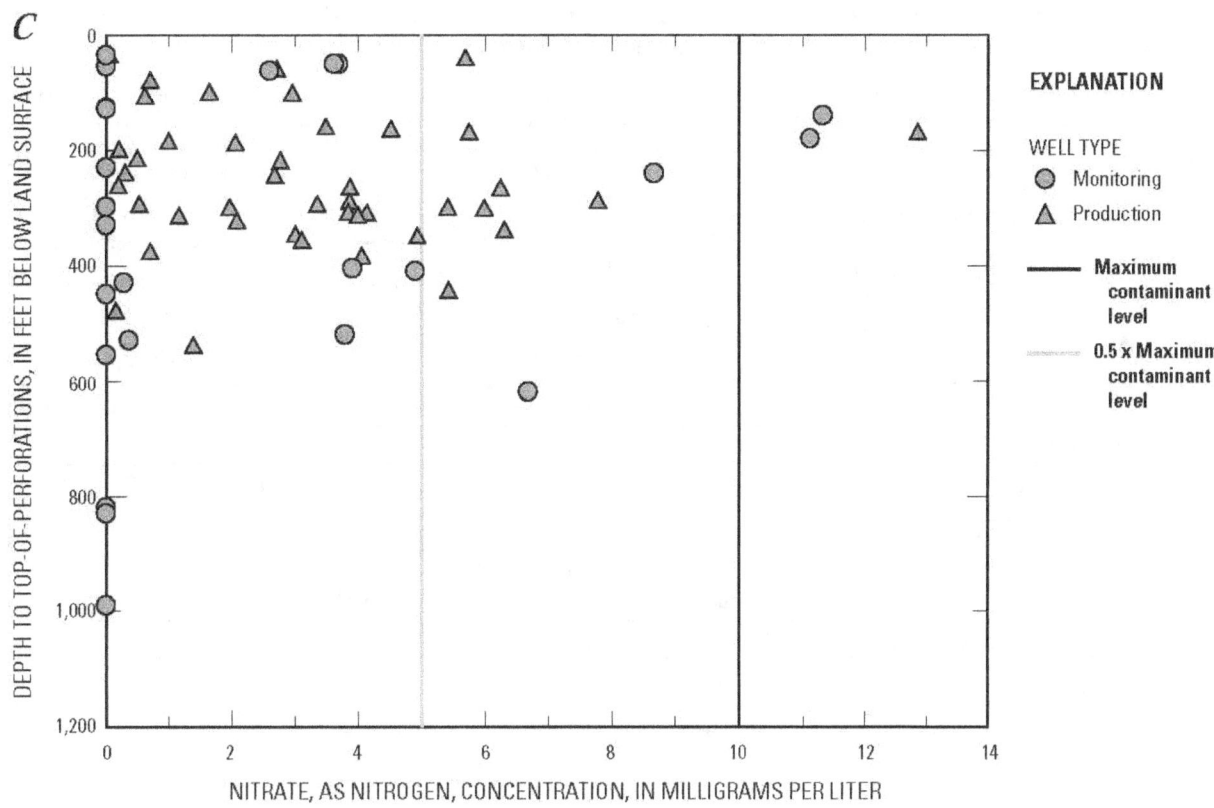

Figure 14.—Continued.

Table 10.   Results of Wilcoxon rank-sum tests on grid-well data used to determine significant differences between selected water-quality constituents grouped by potential explanatory factor classifications, San Francisco Bay study unit, California GAMA Priority Basin Project.

[Modern, sample with water recharged after 1952; Pre-modern, sample with water recharged before 1952; Mixed age, sample with modern and pre-modern components; mg/L, milligrams per liter; ≥, greater than or equal to; <, less than; ≤, less than or equal to; oxic, dissolved oxygen ≥ 0.5 mg/L; anoxic/suboxic, dissolved oxygen < 0.5 mg/L; mixed redox, more than one redox state; only results with p-values ≤ 0.05 are considered significant in this study; Wilcoxon rank-sum tests with exact distribution and continuity correction; Z, test statistic for Wilcoxon test; significantly positive Z-value (first classification is larger than second); significantly negative Z-value (first classification is smaller than the second); nd, no significant difference (p-value > 0.05); CFC-113, 1,1,2-Trichlorotrifluoroethane]

	Potential explanatory factors					
	Groundwater age classification			Redox classifications		
Selected water-quality constituents	Modern age compared with mixed age	Mixed age compared with pre-modern age	Modern age compared with pre-modern age	Oxic conditions compared with mixed conditions	Mixed conditions compared with anoxic/suboxic conditions	Oxic conditions compared with anoxic/suboxic conditions
	Z: Test statistic for Wilcoxon test			Z: Test statistic for Wilcoxon test		
Iron	nd	nd	nd	−3.54	2.55	nd
Manganese	nd	nd	nd	−2.99	nd	−2.27
Total dissolved solids	nd	nd	nd	nd	nd	nd
Chloride	nd	nd	nd	nd	nd	nd
Nitrate	nd	2.71	2.62	nd	2.60	3.17
Chloroform	nd	nd	nd	nd	nd	nd
Solvents, sum of concentrations	2.16	nd	nd	nd	2.93	2.01
CFC-113	nd	nd	2.18	nd	2.00	2.01
Perchlorate	nd	2.76	nd	nd	nd	2.24

Table 11. Results of non-parametric (Spearman's method) analysis of correlations between selected water-quality constituents and potential explanatory factors, San Francisco Bay study unit, California GAMA Priority Basin Project.

[Only results with p-values ≤ 0.05 are considered significant in this study. A positive value indicates postive correlations; negative values indicate negative correlations. ρ, Spearman's correlation statistic; >, greater than; ≥, greater than or equal to; nc, no significant correlation (p-value > 0.05); TDS, total dissolved solids; USEPA, U.S. Environmental Protection Agency; CDPH, California Department of Public Health; CFC-113, 1,1,2-Trichlorotrifluoroethane; MCL-US, USEPA maximum contaminant level; MCL-CA, CDPH maximum contaminant level; SMCL-CA, CDPH secondary maximum contaminant level]

Constituent	Benchmark type	High aquifer proportion, percent	Data analyzed: USGS-grid and -understanding production wells				Data analyzed: USGS-grid wells		
			Well depth	Depth to top-of-perforations	pH	Dissolved oxygen (DO)	Percent urban land use[1]	Percent agricultural land use[1]	Percent natural land use[1]
			ρ: Spearman's correlation statistic				ρ: Spearman's correlation statistic		
Inorganic constituents									
Iron[2]	SMCL-CA	3.0	nc	nc	nc	nc	nc	nc	nc
Manganese[2]	SMCL-CA	12	nc	nc	nc	nc	nc	nc	nc
Nitrate[2]	MCL-US	2.1	nc	nc	nc	0.49	0.29	nc	−0.29
TDS[2]	SMCL-CA	7.0	nc	nc	nc	−0.31	nc	nc	nc
Chloride[2]	SMCL-CA	6.1	nc	nc	−0.49	nc	nc	−0.31	nc
Organic constituents and constituent classes									
Chloroform[3]	MCL-US	0.0	nc	nc	nc	nc	nc	nc	nc
Solvents, sum of concentrations[3]	variable	0.3	nc	nc	nc	nc	nc	−0.25	nc
CFC-113[3]	MCL-CA	0.0	0.31	nc	nc	0.33	0.33	nc	−0.33
Constituents of special interest									
Perchlorate[3]	MCL-US	0.0	nc	nc	nc	0.27	nc	nc	−0.24

[1] Land-use percentages are within a radius of 500 meters centered around each well included in analysis.

[2] Constituents with ≥ 2% high aquifer proportion.

[3] Classes of compounds that include constituents with high and (or) moderate values or detection frequencies at any concentration ≥ 10%.

Constituents with Aesthetic Benchmarks

Eight inorganic constituents have SMCL-CA benchmarks that are based on aesthetic properties. Four of these—chloride, TDS, iron, and manganese—met the criteria for additional evaluation (table 7). Relative-concentrations of TDS were high in 7.0 percent of the primary aquifer system, and were moderate in 37 percent (table 7). Chloride was detected at a high relative-concentration in 6.1 percent of the primary aquifer system. Iron had a high relative-concentration in 3.0 percent and a moderate relative-concentration in 6.1 percent of the primary aquifer system (table 7; figs. 11, 12). Manganese had a high relative-concentration in 12 percent and a moderate relative-concentration in 6.1 percent of the primary aquifer system.

Understanding Assessment for TDS and Chloride

Natural sources of TDS include mixing of groundwater with deep saline groundwater that is influenced by interactions with deep marine or lacustrine sediments, saltwater intrusion, and water-rock interactions in older waters. Potential anthropogenic sources of TDS to groundwater in the SFBAY study unit include urban irrigation, evaporation, wastewater disposal and industrial effluent, artificial recharge, and leaking sewer pipes.

The anion chloride is a major component of TDS in the SFBAY study unit, and its distribution, for the most part, reflects that of TDS. TDS is negatively correlated with DO (table 11; fig. 15). Chloride concentrations are negatively correlated with pH and percentage of agricultural land use (table 11), suggesting that the source of high chloride and high TDS waters is not from agricultural irrigation return. The moderate and high TDS grid wells are located in three areas of the study unit (fig. 13B). On the east side of the Bay, high and moderate relative-concentrations of TDS were detected in areas with history of intrusion of water from the Bay in response to pumping of freshwater from aquifers. A combination of drought periods and pumping in the early 1900s allowed the groundwater level to fall below sea level in well fields near the Bay, and saltwater intruded into the shallow aquifers (Figuers, 1998). Continuous pumping in the Niles Cone subbasin allowed saline waters to migrate from the shallow aquifers through the Bay Mud to the deeper aquifers used for public supply (Figuers, 1998). On the southwest

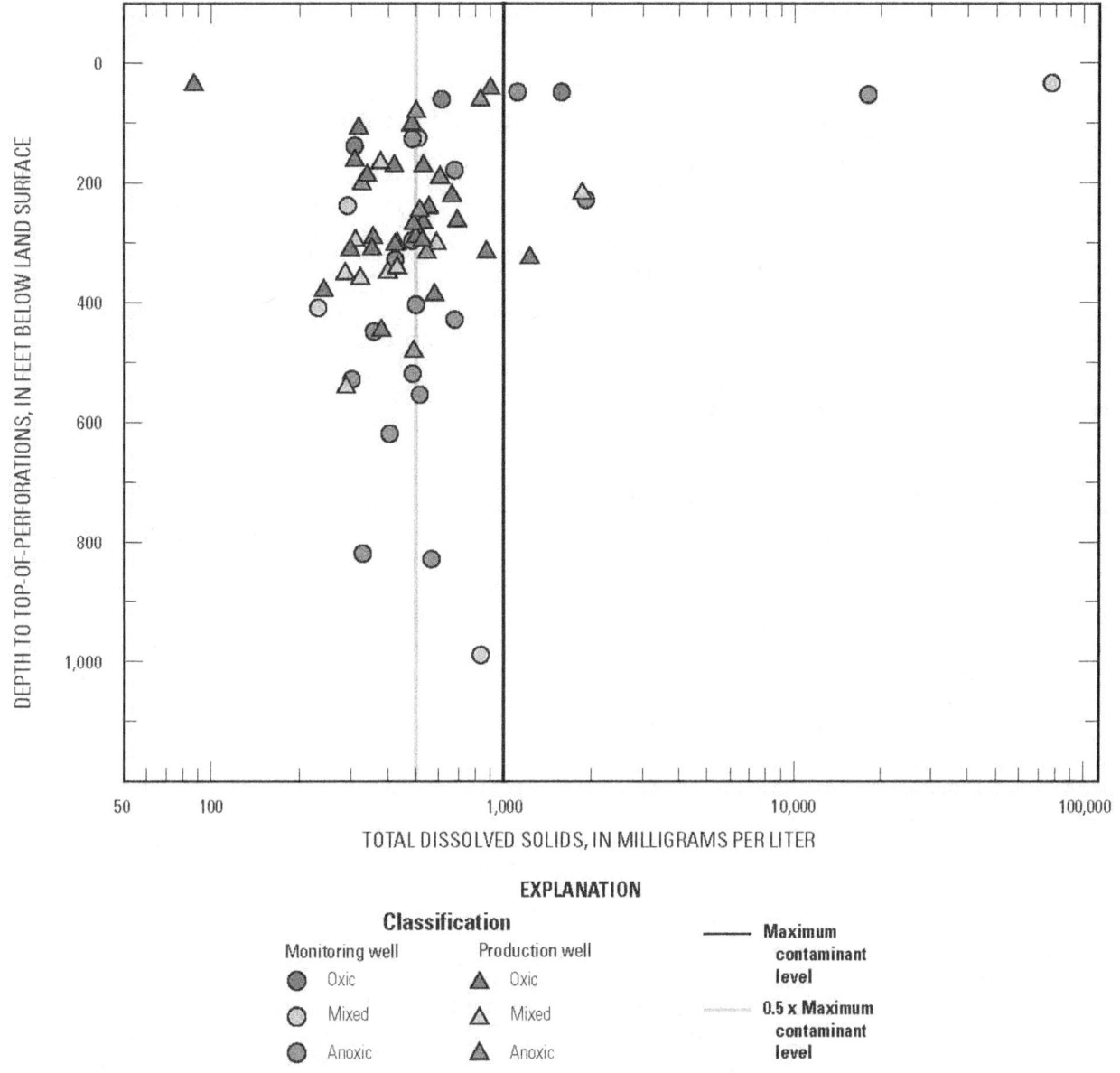

Figure 15. Total dissolved solids concentrations relative to depth to top-of-perforation, well type, and geochemical classifications in USGS-grid and USGS-understanding wells in the San Francisco Bay study unit, California GAMA Priority Basin Project, April–June 2007.

side of the Bay, moderate relative-concentrations of TDS occur on an alluvial fan near the town of Atherton. Upwelling of connate saline water from marine sediments may be a source of TDS to the aquifers (Metzger and Fio, 1997). On the east side of the Santa Clara Valley, moderate relative-concentrations of TDS occur in wells along Coyote Creek, which drains marine sediments in the Diablo Range to the east of the study unit. Coyote Creek flows northwest, through the fan deposits of the Santa Clara Valley, to the Bay. Numerous

impoundments in the southern part of the valley store imported water and then release the water into Coyote Creek for groundwater recharge. Accumulation of salts in Coyote Creek along the flowpath from both of these waters may be a source of TDS to the aquifers.

Many monitoring wells in the SFBAY study unit were selected for sampling in order to collect data in areas with known salinity issues. The SFM-B cluster is dedicated to monitoring seawater intrusion to aquifers on the west side of the Bay. The SFM-D and SFM-E clusters are in areas of the East Bay where brackish waters are pumped for desalination plants. High relative-concentrations of TDS and (or) chloride were detected in these monitoring well clusters, in addition to the shallowest well of the SFM-F cluster in the East Bay. The highest concentrations of TDS were detected in wells SFM-B2 (18,200 mg/L) and SFM-F6 (77,800 mg/L) and are consistent with observations of brackish to hypersaline waters in the shallow aquifers near the San Francisco Bay (California Environmental Protection Agency, San Francisco Bay Regional Water Quality Control Board, 2003). As noted previously, continuous pumping in the Niles Cone subbasin of the Santa Clara Valley (East Bay) also has allowed bay water and saline water from adjacent salt ponds to enter the shallow and deeper aquifers (Figuers, 1998; California Environmental Protection Agency, San Francisco Bay Regional Water Quality Control Board, 2003).

Understanding Assessment for Iron and Manganese

Potential natural sources of iron and manganese to groundwater include the dissolution of iron and manganese oxyhydroxide minerals present in many sediments and rocks, and of primary iron- and manganese-bearing silicate minerals found in igneous and metamorphic rocks (Hem, 1970). Potential anthropogenic sources of these constituents to groundwater include effluents associated with the steel and mining industries (Reimann and de Caritat, 1998) and soil amendments, in the form of manganese and iron sulfates, that are added to deficient soils in order to stimulate crop growth. Concentrations of iron and manganese are strongly influenced by redox conditions in the aquifer. In sediments, iron and manganese oxyhydroxide minerals are common as suspended particles and as coatings on mineral surfaces (Sparks, 1995). These minerals are stable in oxygenated systems at neutral pH. Under anoxic conditions, however, the process of reductive dissolution can mobilize iron and manganese cations from oxyhydroxides, increasing their concentrations in groundwater (Sparks, 1995). Oxidation-reduction conditions for the SFBAY study unit are reported in appendix D (table D3).

In the SFBAY study unit, concentrations of iron in production wells were significantly lower in oxic groundwater than in mixed groundwater (table 10). Concentrations of manganese were also significantly lower in oxic compared with mixed or anoxic groundwater (table 10). These results are consistent with reductive dissolution of iron and manganese oxyhydroxides in the aquifer sediments, which takes place in the downgradient parts of the aquifer system.

Fourteen monitor wells (from five well clusters) had high relative-concentrations of iron and (or) manganese. As discussed previously, most monitor wells have waters classified as anoxic and classified as pre-modern age and are at the distal end of the groundwater flow path. Elevated

concentrations of iron and (or) manganese in these wells would also be consistent with reductive dissolution of iron and manganese oxyhydroxides in the aquifer sediments.

Organic Constituents

The organic compounds are organized by constituent class, including volatile organic compounds (VOCs) and pesticides. VOCs are found in paints, solvents, fuels, and refrigerants; they can be byproducts of water disinfection and are characterized by their volatile nature, or tendency to evaporate. Pesticides include herbicides, insecticides, and fungicides and are used to control weeds, insects, or fungi in agricultural, urban, and suburban settings.

The proportion of the aquifer with high relative-concentrations of organic constituents with health-based benchmarks was 0.6 percent (spatially weighted value) (table 9), resulting from high spatially weighted relative-concentrations of the solvent tetrachloroethene (PCE) (0.3 percent) and the gasoline additive methyl *tert*-butyl ether (MTBE) (0.3 percent) (table 7). The proportion of the aquifer having moderate relative-concentrations of organic constituents was 12 percent (table 9). Only 30 of the 202 organic compounds analyzed for were detected, and most of these detected constituents (23 of the 30) have health-based benchmarks (table 8).

VOCs were detected in 49 percent of the primary aquifer system. Of the 14 VOCs detected, 10 were detected only at low relative-concentrations. The maximum relative-concentrations of four VOCs—PCE, trichloroethene (TCE), chloroform, and 1,1-dichloroethene—were greater than 0.1 but less than 1.0 (figs. 16, 17). Three VOCs—chloroform, 1,1,1-trichloroethane (TCA), and 1,1,2-trichlorotrifluoroethane (CFC-113)—were detected in more than 10 percent of the grid wells (figs. 17, 18). All concentrations of VOCs in samples from USGS-grid and understanding wells were less than health-based benchmarks.

Pesticides or pesticide degradates were detected in 19 percent of the primary aquifer system. Of the 117 pesticides and pesticide degradates analyzed for, 6 were detected in the USGS-grid wells; 3 were pesticide parent compounds with benchmarks, 2 were parent compounds without a benchmark, and 1 was a degradate without a benchmark. All concentrations of pesticides were below health-based benchmarks. One herbicide, atrazine, was detected at moderate relative-concentration in one sample (table 7; fig. 17). The herbicide degradate deethylatrazine was the only pesticide detected in 10 percent or more of the grid wells. Deethylatrazine does not have a benchmark, and therefore is not included in the status or understanding assessments. The individual constituents that were not detected and the wells sampled in the SFBAY study unit are listed in Ray and others (2009).

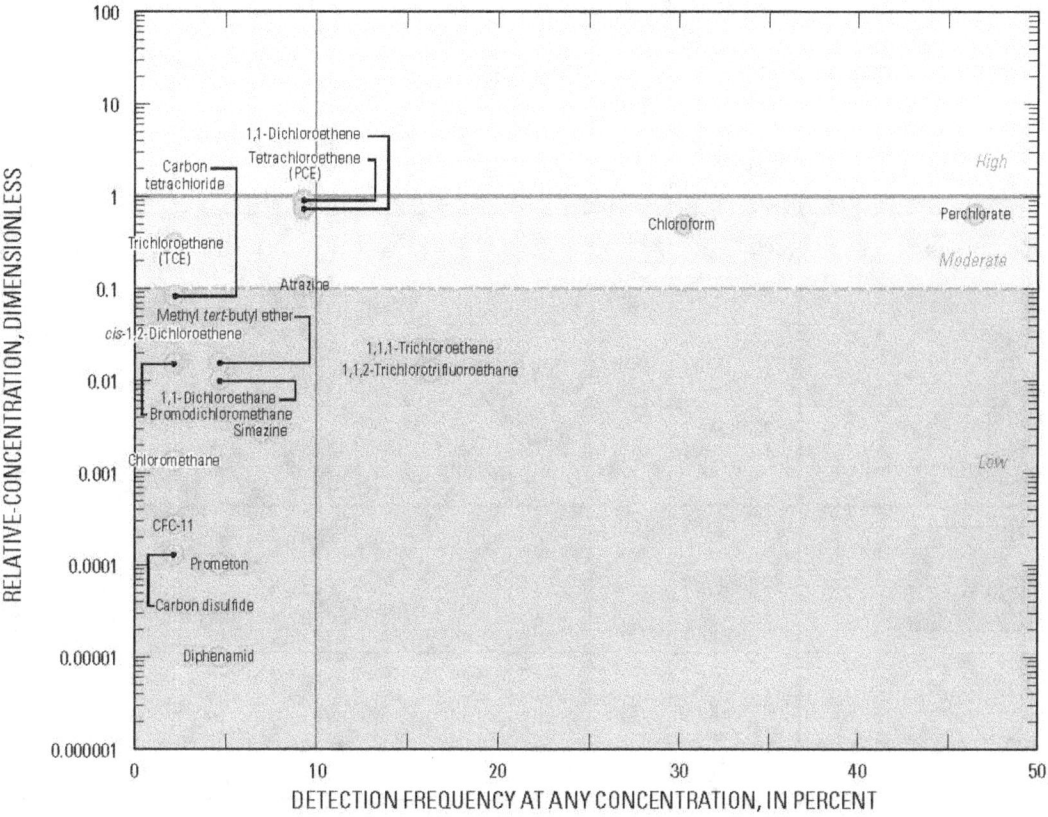

EXPLANATION

Prometon   **Name and center of symbol is the maximum relative-concentration for that constituent—**
Unless indicated by following location line: ⌐   (more than 25 grid wells sampled)

Figure 16.   Detection frequency and maximum relative-concentrations of organic and special-interest constituents detected in USGS-grid wells in the San Francisco Bay study unit, California GAMA Priority Basin Project, April–June 2007.

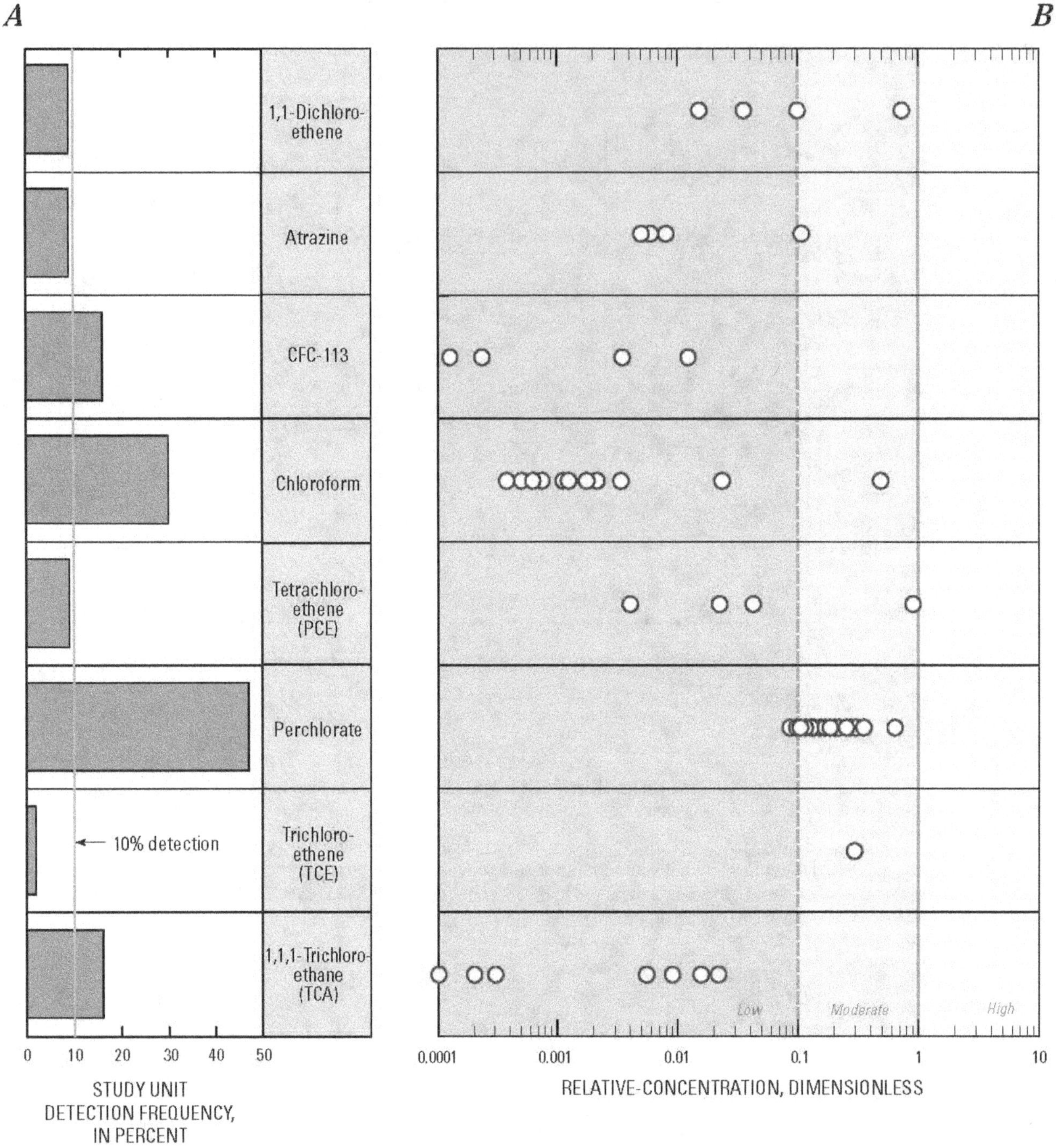

Figure 17. (A) Detection frequency and (B) maximum relative-concentrations of selected organic and special-interest constituents in USGS-grid wells in the San Francisco Bay Study Unit, California GAMA Priority Basin Project, April–June 2007.

A

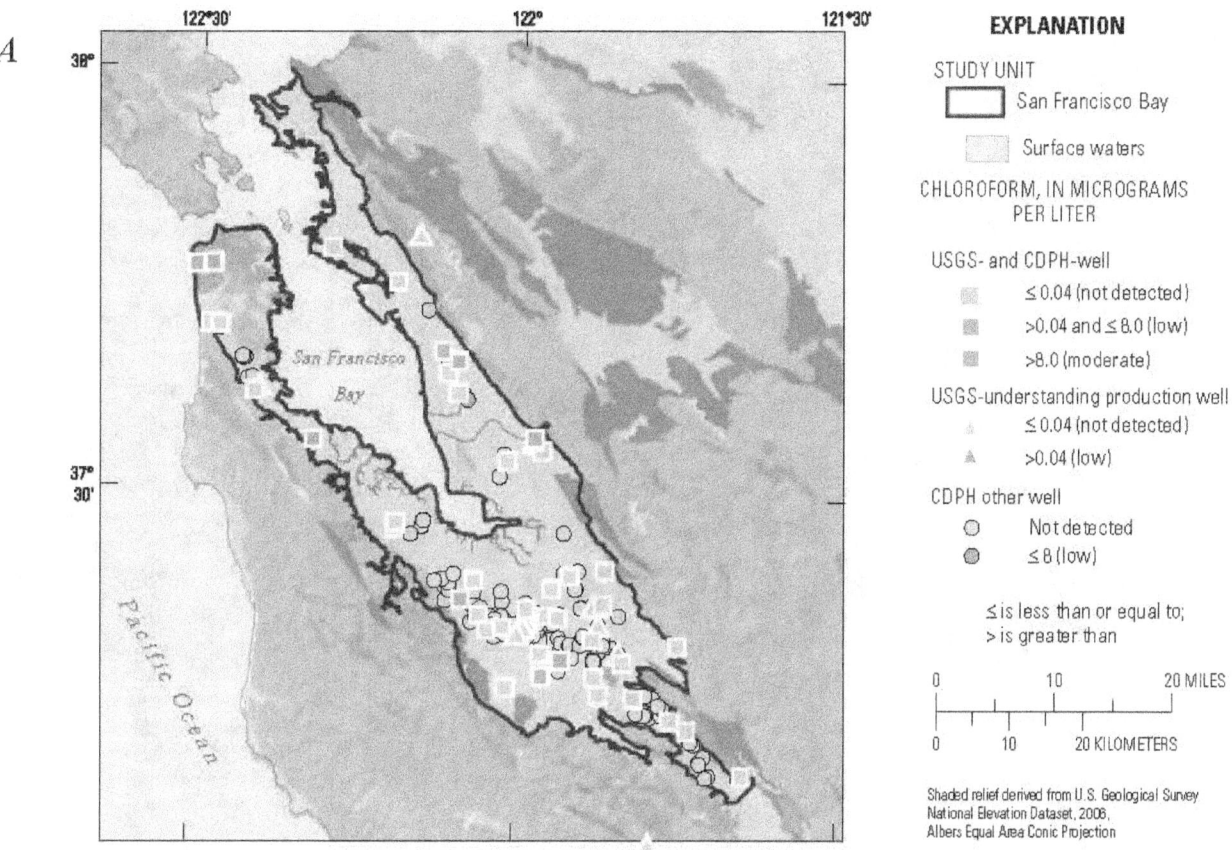

EXPLANATION

STUDY UNIT

☐ San Francisco Bay

▨ Surface waters

CHLOROFORM, IN MICROGRAMS
PER LITER

USGS- and CDPH-well

▨ ≤ 0.04 (not detected)

▨ >0.04 and ≤ 8.0 (low)

▨ >8.0 (moderate)

USGS-understanding production well

▲ ≤ 0.04 (not detected)

▲ >0.04 (low)

CDPH other well

○ Not detected

◉ ≤ 8 (low)

≤ is less than or equal to;
> is greater than

Shaded relief derived from U.S. Geological Survey
National Elevation Dataset, 2006,
Albers Equal Area Conic Projection

Figure 18. Concentrations of selected organic and special-interest constituents detected in USGS-grid and USGS-understanding wells for April–June 2007, San Francisco Bay study unit, California GAMA Priority Basin Project: (*A*) chloroform, (*B*) solvents, (*C*) atrazine, and (*D*) perchlorate.

B

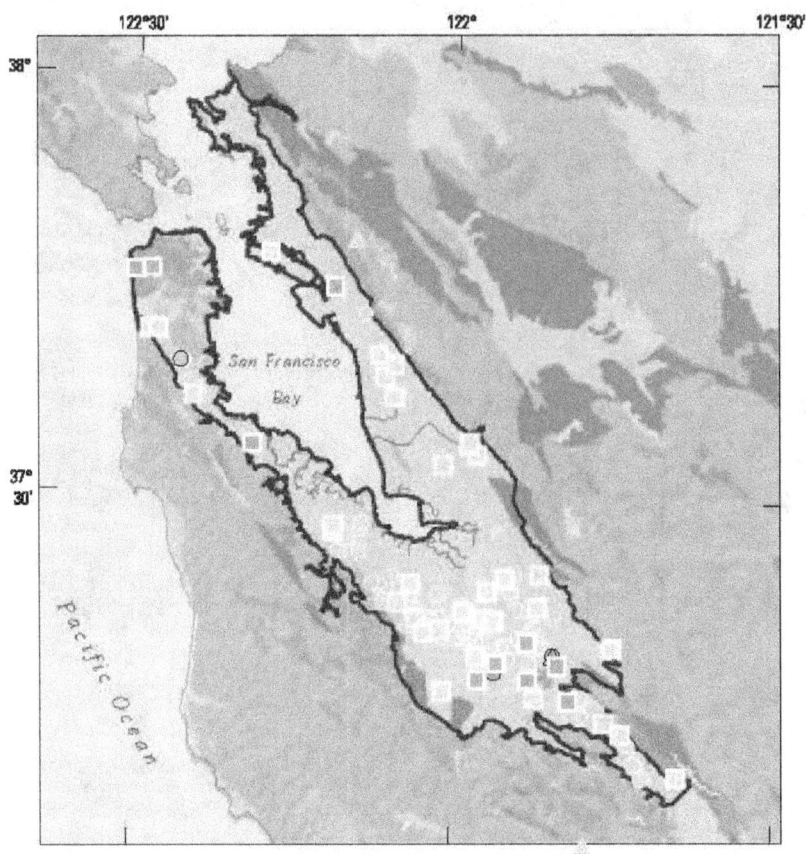

Shaded relief derived from U.S. Geological Survey
National Elevation Dataset, 2006.
Albers Equal Area Conic Projection

EXPLANATION

STUDY UNIT

☐ San Francisco Bay

▨ Surface waters

SUM OF SOLVENTS,
IN MICROGRAMS PER LITER

USGS- and CDPH-well

▨ < 0.02 (not detected)

▨ 0.02 to 2.5 (low)

▨ 2.6 to 5.0 (moderate)

USGS-understanding production well

△ < 0.03 (not detected)

△ 0.03 to 0.51 (low)

CDPH other well

○ < 0.03 (not detected)

◉ 0.03 to 2.5 (low)

◉ 2.6 to 5.0 (moderate)

○ >5.1 to 6.57 (high)

< is less than

Concentration is the sum of the concentrations of tetrachloroethene (PCE), trichloroethene (TCE), 1,1,1-trichloroethane (TCA), 1,1-dichloroethane, *cis*-1,2-dichloroetheylene, and tetrachloromethane.

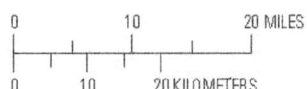

Figure 18.—Continued.

C

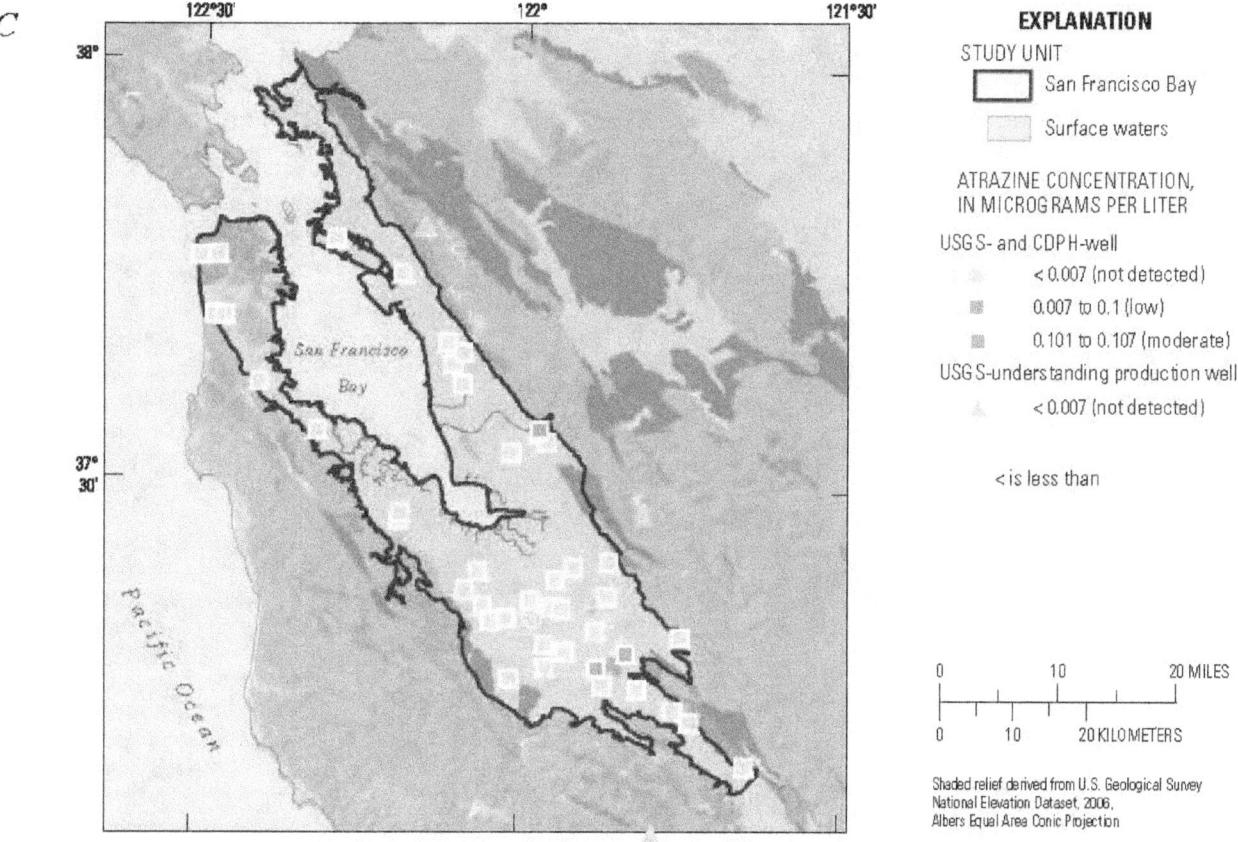

EXPLANATION

STUDY UNIT

☐ San Francisco Bay

☐ Surface waters

ATRAZINE CONCENTRATION,
IN MICROGRAMS PER LITER

USGS- and CDPH-well

   < 0.007 (not detected)

   0.007 to 0.1 (low)

   0.101 to 0.107 (moderate)

USGS-understanding production well

   < 0.007 (not detected)

   < is less than

0   10   20 MILES

0   10   20 KILOMETERS

Shaded relief derived from U.S. Geological Survey
National Elevation Dataset, 2006,
Albers Equal Area Conic Projection

Figure 18.—Continued.

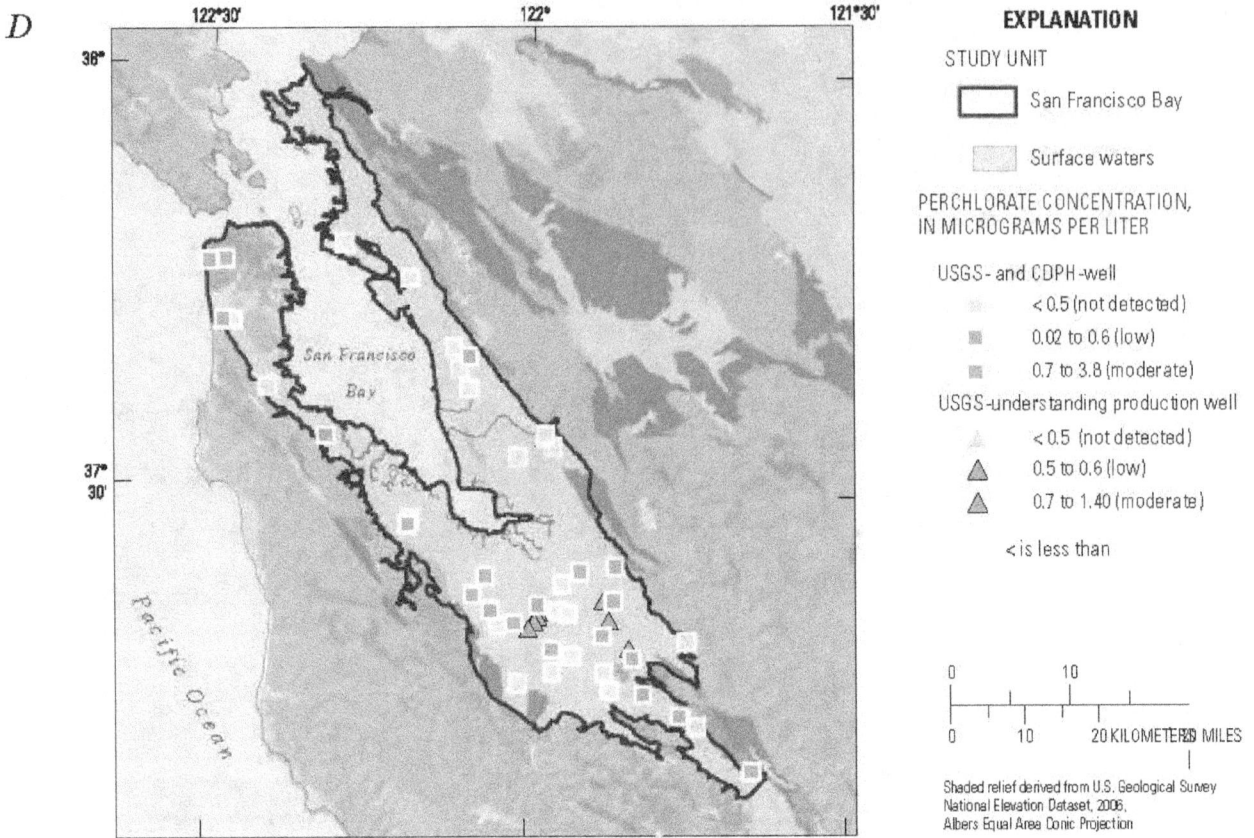

Figure 18.—Continued.

Trihalomethanes

Chloroform was detected in 30 percent of the 43 grid wells in the study unit (figs. 16, 17) and was detected at low relative-concentrations in 28 percent of the primary aquifer system and at moderate relative-concentrations in 2.3 percent of the primary aquifer system (table 7). Chloroform also was the most frequently detected VOC in groundwater according to the USGS National Water-Quality Assessment (NAWQA) Program (Zogorski and others, 2006). Figure 18A shows the distribution of chloroform in the SFBAY study unit.

Understanding Assessment for Chloroform

Chloroform concentrations in production wells were not significantly correlated with any explanatory factor (tables 10, 11). Nationally, THM concentrations have been positively correlated with percentage of urban land use (Zogorski and others, 2006). Potential urban sources of THMs include recharge from landscape irrigation that uses disinfected water, leakage from distribution or sewer systems, and industrial and commercial sources (Ivahnenko and Barbash, 2004).

Five of the 14 chloroform detections in production wells occurred in groundwater with mixed ages, 6 detections were in groundwater classified as modern age, and 3 detections were in groundwater classified as pre-modern age. Because water supplies have been disinfected with chlorine during the past 100 years, chloroform may be in wells that do not appear to have modern recharge (since 1952) water.

Solvents

Solvents are used for various industrial, commercial, and domestic purposes. One solvent compound met the selection criteria of greater than 10 percent detection frequency; 1,1,1-trichloroethane (TCA) was detected in 16 percent of the grid wells. Two solvents, PCE and TCE, each had moderate aquifer proportions of 2.3 percent (table 7). Solvents had a high aquifer-scale proportion of 0.3 percent (spatially weighted), reflecting detections of PCE (tables 7, 9). The proportion of the aquifer having moderate values of solvents was 4.7 percent.

Historically high concentrations of the solvent 1,1-dichloroethane were recorded in the CDPH database for the period before April 1, 2004, but high concentrations were not detected during the current period of study (table 4).

Understanding Assessment for Solvents

The sum of solvent concentrations was calculated by summing the concentrations of all six solvents detected: PCE, TCE, TCA, 1,1-dichloroethane, cis-1,2-dichloroethylene, and carbon tetrachloride. The sum of solvents in production wells was negatively correlated with percentage of agricultural land use (table 11). Nationally, solvent concentrations also have been correlated strongly with percentage of urban land use (Zogorski and others, 2006; Moran and others, 2007). Figure 18B shows the distribution of solvents in the SFBAY study unit.

Solvent concentrations in production wells were significantly greater in modern compared with mixed-age waters. Some solvents that were used before 1952 could be present in pre-modern or mixed-age water. Solvent concentrations in production wells were significantly greater in oxic or mixed redox waters compared with anoxic waters. This reflects the relation between DO concentrations and groundwater age (table 6).

Historical releases of chlorinated solvents to the groundwater aquifers in the SFBAY study unit are related to electronics manufacturing in the Silicon Valley and dry cleaner locations in Santa Clara Valley (Santa Clara Valley Water District, 2011). However, these releases are typically above the confined zone of the aquifer and do not reach the deeper parts of the aquifer used for public supply. Thus, similar to previous work in the Bay area (Moran and others, 2002), solvent occurrences in groundwater samples cannot be traced to known solvent plumes.

One monitoring well had a detection of solvents, reflecting a detection of carbon tetrachloride. This was the shallowest well (depth to top of perforations of 140 ft) of a cluster in the northwestern portion of the study unit, near San Francisco (SFM-A cluster). Similar to production wells with solvent detections, this monitor well had oxic, mixed-age groundwater.

Other VOCs

Other VOCs were present at high relative-concentrations in 0.3 percent of the primary aquifer system and at moderate relative-concentrations in 2.3 percent (table 9). The gasoline oxygenate methyl tert-butyl ether was present at high relative-concentration in 0.3 percent (spatially weighted) of the primary aquifer system (table 7). The VOC 1,1,2-trichlorotrifluoroethane (CFC-113)—often used as a refrigerant—was detected in 16 percent of the grid wells (fig. 16). Refrigerants are among the most frequently detected VOCs in the nation (Zogorski and others, 2006).

The fumigant 1,2-dichloropropane had historically high relative-concentrations in the CDPH database from the period before April 1, 2004, but not during the current period of study (table 4).

Understanding Assessment for CFC-113

Concentrations of CFC-113 in production wells were significantly greater in modern-age waters compared with pre-modern waters and in oxic and mixed redox waters compared with anoxic/suboxic waters (table 10; fig. 19). CFC-113 was positively correlated with DO concentrations and negatively correlated with well depth (table 11). CFC-113 was positively correlated with percentage of urban land use and equally negatively correlated with percentage of natural land use, as is expected because of the anthropogenic source of the compound. Some of the explanatory variables related to CFC-113, such as DO and groundwater age, are generally related to each other (table 6). There were no detections of CFC-113 in any monitoring well.

Herbicides

Herbicides were not detected at high relative-concentrations in the SFBAY study unit and were detected at moderate relative-concentrations in 2.3 percent of the primary aquifer system (table 9). Atrazine was detected in 9.3 percent of the grid wells and was the only herbicide detected at moderate relative-concentrations (figs. 16, 18C; table 7). Atrazine was among the most commonly detected herbicides in groundwater in major aquifers across the United States (Gilliom and others, 2006).

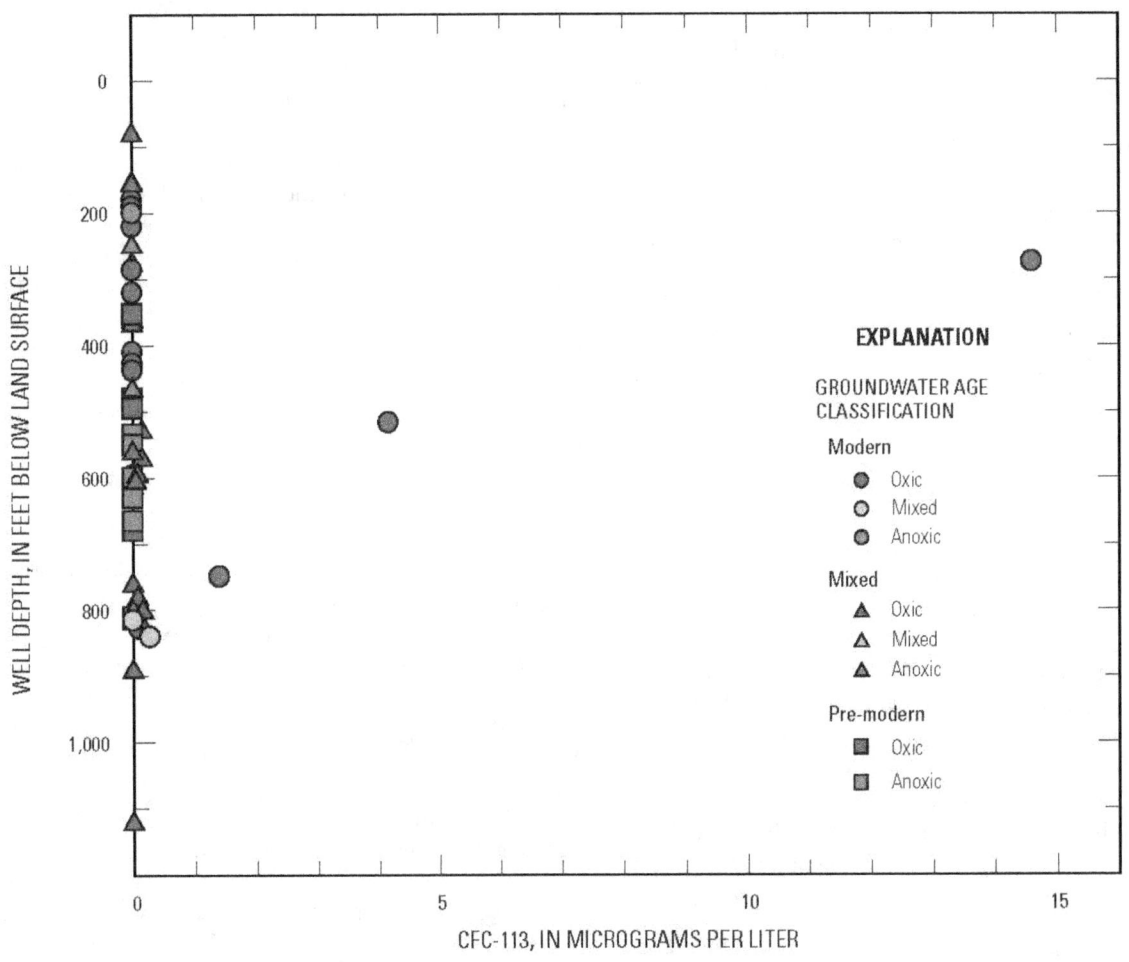

Figure 19. Concentrations of CFC-113 relative to well depth and groundwater age classification and geochemical classification for the San Francisco Bay study unit, California GAMA Priority Basin Project, April–June 2007.

Special-Interest Constituents

Constituents of special interest analyzed for in the SFBAY study unit were *N*-nitrosodimethylamine (NDMA) and perchlorate. These constituents were selected because they recently have been detected in drinking-water supplies or are considered to have the potential to reach drinking-water supplies (California Department of Public Health, 2008a,b,c). NDMA was not detected in the three grid wells sampled (Ray and others, 2009). Perchlorate was analyzed for in samples from all grid wells (table 1). Perchlorate was not detected at high relative-concentrations and was detected at moderate relative-concentrations in 42 percent of the primary aquifer system (table 7; fig. 18*D*).

Understanding Assessment for Perchlorate

Perchlorate concentrations in production wells were negatively correlated with percentage of natural land use (table 11). Concentrations of perchlorate were significantly higher in mixed-age waters than in pre-modern water (table 10; fig. 20). Perchlorate concentrations were positively correlated with DO concentrations (table 11), and concentrations of perchlorate were significantly higher in oxic waters compared with anoxic/suboxic waters (table 10; fig. 21). While perchlorate biodegrades under anoxic conditions in some aquifers (Sturchio and others, 2007), the apparent relation between perchlorate and DO in the San Francisco Bay study unit may result from relations of DO and perchlorate with modern and mixed-age groundwater.

Perchlorate has natural and anthropogenic sources to groundwater. Perchlorate forms naturally in the atmosphere and is present dissolved in precipitation and deposited in unsaturated zones (Rajagopalan and others, 2006, 2009). The distribution of perchlorate under natural conditions in groundwater is likely correlated with climate because the extent of evaporative concentration of precipitation in the hydrologic cycle is likely to increase with increasing aridity (Fram and Belitz, 2011). Potential anthropogenic sources of perchlorate include industrial, manufacturing, or commercial uses, such as rocket fuel, explosives, road flares, automobile air-bag systems, and other products, and nitrate fertilizers mined from the Atacama Desert of Chile that have been used historically on some orchard crops (U.S. Environmental Protection Agency, 2005; Dasgupta and others, 2006; Bohlke and others, 2009). In addition, irrigation may remobilize naturally deposited perchlorate salts in the unsaturated zone. Concentrations greater than 1 µg/L (relative-concentration of 0.17) have a high probability of being anthropogenic in origin (Fram and Belitz, 2011).

Fram and Belitz (2011) used logistic regression to quantify the relation between the probability of perchlorate detection in groundwater resources sampled by the GAMA Priority Basin Project and water-quality indicators and other proxies of natural and anthropogenic perchlorate sources and processes. Their best-fit model was constructed with aridity index as the proxy for perchlorate under natural conditions and with an Anthropogenic Score (AS) as the proxy for anthropogenic sources and processes affecting perchlorate. The AS summed four binary categorical variables: presence or absence of herbicides and fumigants, presence or absence of solvents and fuel components, nitrate concentration above or below 3 mg/L, and presence or absence of known sites of perchlorate contamination within a specific distance of the well. This model was applied to the data from the SFBAY study unit to assess whether the observed concentrations and detection frequencies of perchlorate can be accounted for by natural conditions or require additional inputs of perchlorate from anthropogenic sources or processes.

USGS production wells from the SFBAY study unit were divided into four groups having AS values of 0, 1, 2, and 3 or 4. The predicted probabilities of detecting perchlorate under natural conditions (AS=0) at concentrations greater than 0.5 µg/L and greater than 1.0 µg/L are approximately 8.1 percent and 2.7 percent, respectively, at the average aridity index for USGS production well samples (0.37) (fig. 22*A*, *B*; Fram and Belitz, 2011). The predicted probabilities increase as AS increases, reaching 80 percent and 66 percent, respectively, when AS=4. The detection frequencies observed in samples from the SFBAY study unit also increase as AS increases and are generally higher than the predicted probabilities at each value of AS (fig. 22*A*, *B*).

The fact that the observed detection frequencies are greater than the predicted probabilities of detection indicates that the observed distribution of perchlorate in the SFBAY study unit is not representative of the distribution expected under natural conditions, and indicates that anthropogenic sources of perchlorate are dominant. This was expected, given that there are several known sites of perchlorate contamination to groundwater from industrial sources in the study unit (U.S. Environmental Protection Agency, 2005).

Five monitoring wells (from three well clusters) had detections of perchlorate, all at moderate relative-concentrations. All monitoring wells with perchlorate detections were classified as mixed-age groundwaters. Three wells (two well clusters, SFM-C and SFM-E) were classified as anoxic, and one well each (from the same cluster, SFM-A) was classified as mixed and oxic. The two wells from the SFM-A cluster that had detections of perchlorate also had detections of other organic compounds—chloroform and solvents. The other wells with perchlorate detections did not have detections of other organic compounds.

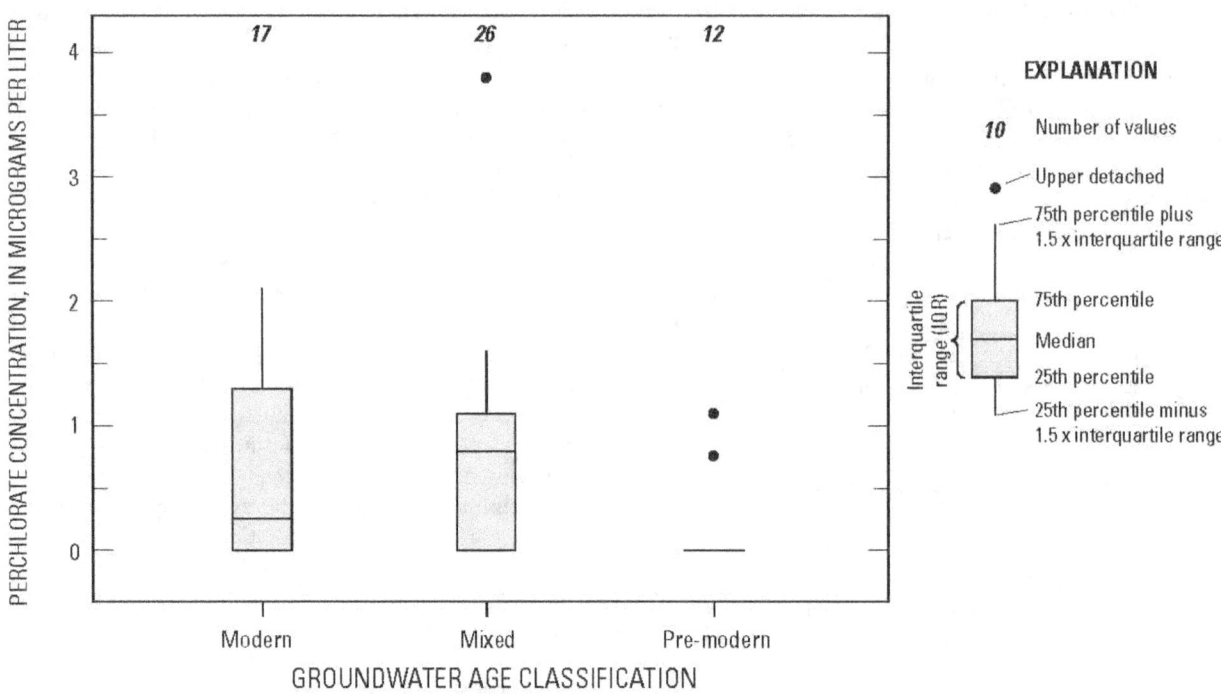

Figure 20.   Concentrations of perchlorate relative to groundwater age classification for the San Francisco Bay study unit, California GAMA Priority Basin Project, April–June 2007.

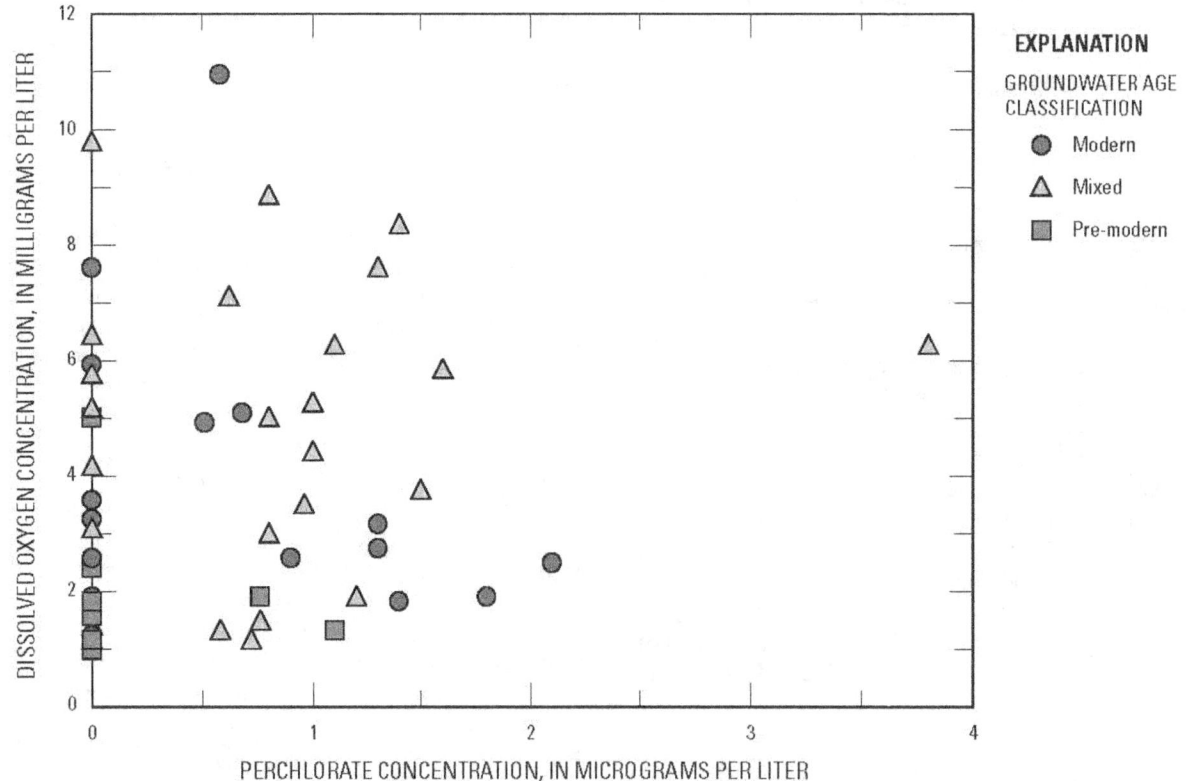

Figure 21.   Concentrations of perchlorate relative to dissolved oxygen concentrations, land-use classification, and groundwater age classification for the San Francisco Bay study unit, California GAMA Priority Basin Project, April–June 2007.

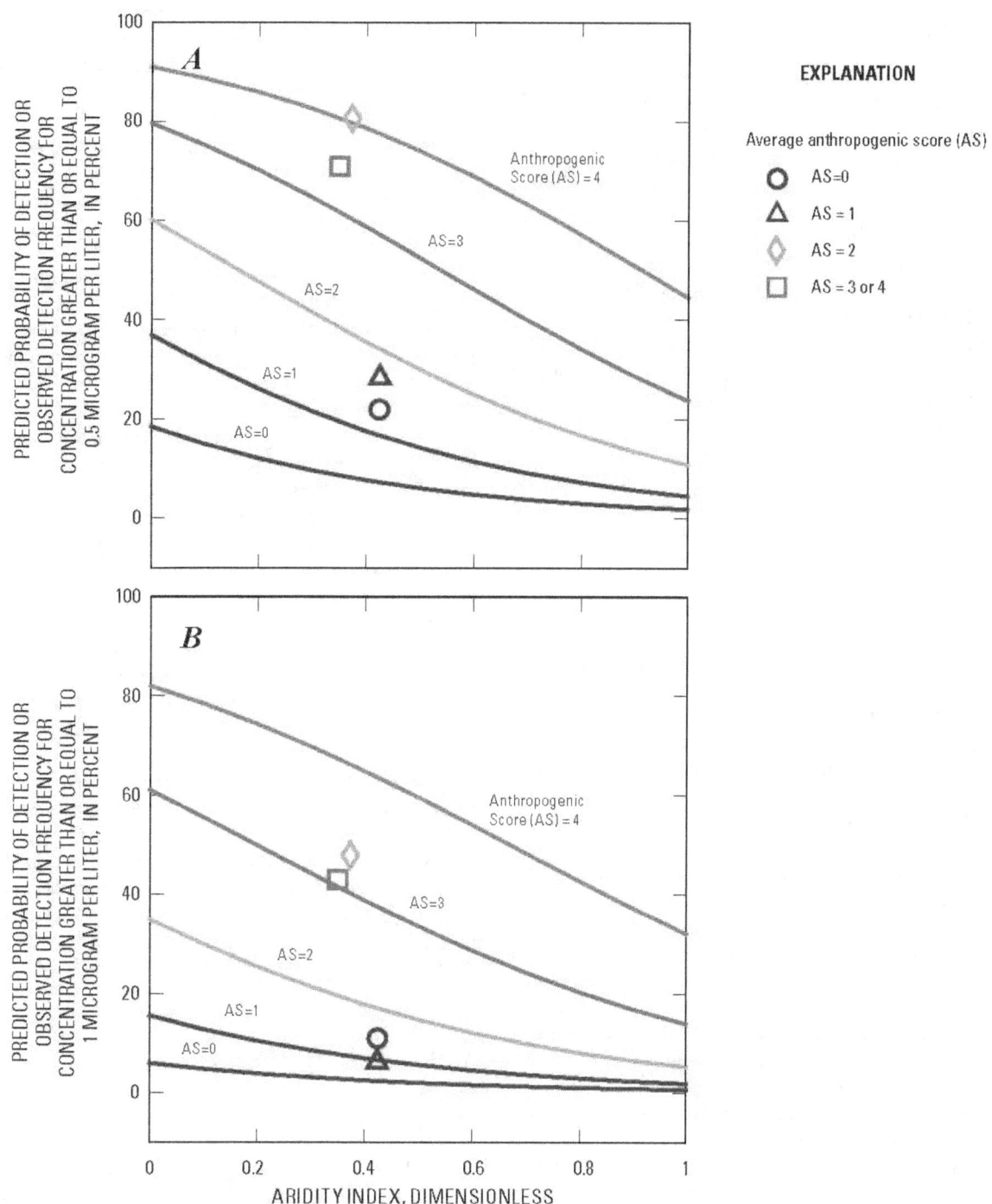

Figure 22. Predicted probability of detecting perchlorate in groundwater as a function of aridity index and Anthropogenic Score and observed detection frequency and average aridity index in groups of samples for perchlorate concentrations greater than or equal to (*A*) 0.5 microgram per liter (µg/L) and (*B*) 1.0 µg/L, San Francisco Bay study unit, California GAMA Priority Basin Project.

Summary

Groundwater quality in the approximately 620-square-mile San Francisco Bay study unit was investigated as part of the Priority Basin Project of the Groundwater Ambient Monitoring and Assessment (GAMA) Program. The GAMA San Francisco Bay (SFBAY) study was designed to provide a spatially unbiased characterization of the quality of untreated groundwater in the primary aquifer system at the basin-scale. The assessment is based on water-quality and ancillary data collected by the U.S. Geological Survey (USGS) from 79 wells in 2007 and water-quality data from the California Department of Public Health (CDPH) database.

The first component of this study, the status of the quality of the current groundwater resource, was assessed by using data from samples analyzed for inorganic constituents, such as major ions and trace elements, volatile organic compounds (VOCs), and pesticides. The status assessment characterizes the quality of groundwater resources within the primary aquifer system of the SFBAY study unit, not the drinking water delivered to consumers by water purveyors.

Relative-concentrations (sample concentration divided by the benchmark concentration) were used to evaluate groundwater quality for those constituents that have Federal and (or) California benchmarks. Aquifer-scale proportion was used as a metric for evaluating regional-scale groundwater quality. High aquifer-scale proportion is defined as the percentage of the primary aquifer system with relative-concentration greater than 1.0 for a particular constituent or class of constituents; proportion is based on an areal rather than a volumetric basis. Moderate and low aquifer-scale proportions were defined as the percentage of the aquifer with moderate and low relative-concentrations, respectively. Two statistical approaches, grid-based and spatially weighted, were used to evaluate aquifer-scale proportion for individual constituents and classes of constituents. Grid-based and spatially weighted estimates were comparable in the SFBAY study unit (90-percent confidence intervals). However, the spatially weighted approach was superior to the grid-based proportion when concentrations of a constituent were high in a small portion of the aquifer.

For inorganic constituents with health-based benchmarks, relative-concentrations were high in 5.1 percent of the primary aquifer system, moderate in 25 percent, and low in 70 percent. The high aquifer-scale proportion of inorganic constituents reflected high values of trace elements (3.0 percent) and nutrients (2.1 percent). Inorganic constituents with high values were barium (3.0 percent) and nitrate (2.1 percent). The inorganic constituents with secondary maximum contaminant levels (total dissolved solids, chloride, manganese, and iron) were high in 14 percent of the primary aquifer system, moderate in 33 percent, and low in 53 percent.

In contrast, relative-concentrations of organic constituents (one or more) were high in 0.6 percent, moderate in 12 percent, and low in 88 percent (not detected in 49 percent) of the primary aquifer system. The high aquifer-scale proportion of organic constituents reflected high aquifer-scale proportions of solvents (0.3 percent spatially weighted) and methyl *tert*-butyl ether (0.3 percent spatially weighted). Of the 204 organic and special-interest constituents analyzed for, 34 were detected. Three organic constituents were frequently detected (in 10 percent or more of samples)—the trihalomethane chloroform and the VOCs 1,1,1-trichloroethane and 1,1,2-trichlorotrifluoroethane (CFC-113)—but none of these three were detected at high relative-concentrations.

In the second component of this report, the understanding assessment, some of the primary natural and human factors that affect groundwater quality were identified through evaluation of correlations between land use, physical characteristics of the wells, geochemical conditions of the aquifer, and relative-concentrations of constituents. Results from these analyses attempt to explain the occurrence and distribution of constituents in the SFBAY study unit.

The understanding assessment indicated that nitrate concentrations were negatively correlated with percentage of natural land use and were greater in groundwater having modern or mixed-age classification compared to groundwater having pre-modern-age classification, suggesting anthropogenic sources for the nitrate. High and moderate relative-concentrations of iron and manganese may be attributed to reductive dissolution of manganese and iron oxides. Concentrations of total dissolved solids were highest in areas of known groundwater salinity concerns. Concentrations and detection frequency of perchlorate were greater than predicted for natural conditions, indicating dominance of anthropogenic sources of perchlorate.

Acknowledgments

The authors thank the following cooperators for their support: the California State Water Resources Control Board, Lawrence Livermore National Laboratory, California Department of Public Health, and California Department of Water Resources. We especially thank the cooperating well owners and water purveyors for their generosity in allowing the USGS to collect samples from their wells. Funding for this work was provided by State of California bonds authorized by Proposition 50 and administered by the State Water Board.

References Cited

Bailey, E.H., Irwin, W.P., and Jones, D.L., 1964, Franciscan and related rocks and their significance in the geology of Western California: California Division of Mines and Geology, Bulletin 183, 177 p., 2 pls.

Belitz, Kenneth, Dubrovsky, N.M., Burow, K.R., Jurgens, B.C., and Johnson, T., 2003, Framework for a groundwater quality monitoring and assessment program for California: U.S. Geological Survey Water-Resources Investigations Report 03-4166, 28 p. (Also available at http://pubs.usgs.gov/wri/wri034166/.)

Belitz, K., Jurgens, B., Landon, M.K., Fram, M.S., and Johnson, T., 2010, Estimation of aquifer-scale proportion using equal-area grids—Assessment of regional-scale groundwater quality: Water Resources Research, v. 46, W11550, 14 p., doi:10.1029/2010WR009321. (Also available at http://www.agu.org/pubs/crossref/2010/2010WR009321.shtml.)

Bohlke, J.K., Hatzonger, P., Sturchio, N.C., Gu, B., Abbene, I.J., and Mroczkowski, S.J., 2009, Atacama perchlorate as an agricultural contaminant in groundwater—Chemical and isotopic evidence from Long Island, New York: Environmental Science & Technology, v. 43, p. 5619–5625.

Brown, L.D., Cai, T.T., and DasGupta, A., 2001, Interval estimation for a binomial proportion: Statistical Science, v. 16, no. 2, p. 101–117. (Also available at http://www.jstor.org/stable/2676784.)

California Department of Public Health, 2008a, Perchlorate in drinking water, accessed July 7, 2008, at http://www.cdph.ca.gov/certlic/drinkingwater/Pages/Perchlorate.aspx.

California Department of Public Health, 2008b, California drinking water—NDMA-related activities, accessed July 7, 2008, at http://www.cdph.ca.gov/certlic/drinkingwater/Pages/NDMA.aspx.

California Department of Public Health, 2008c, 1,2,3-Trichloropropane, accessed July 7, 2008, at http://www.cdph.ca.gov/certlic/drinkingwater/Pages/123TCP.aspx.

California Department of Water Resources, 2003, California's groundwater: California Department of Water Resources Bulletin 118, 246 p. (Also available at http://www.water.ca.gov/groundwater/bulletin118/update2003.cfm.)

California Environmental Protection Agency, 2001, Geographic Environmental Information Management System GeoTracker (GEIMS) Leaking Underground Fuel/Storage Tank database (LUFT) [digital data]: Sacramento, California, California Environmental Protection Agency, State Water Resources Control Board, Division of Water Quality.

California Environmental Protection Agency, San Francisco Bay Regional Water Quality Control Board, 2003, A comprehensive groundwater protection evaluation for South San Francisco Bay basins, accessed June 20, 2007, at http://www.swrcb.ca.gov/rwqcb2/pdfsobaygdwtr/SouthBayReport.pdf. [Accessible as of August 8, 2008, at http://www.waterboards.ca.gov/sanfranciscobay/sobayground.shtml.]

California State Water Resources Control Board, 2003, A comprehensive groundwater quality monitoring program for California: Assembly Bill 99 Report to the Governor and Legislature, March 2003, 100 p.

Chapelle, F.H., 2001, Groundwater microbiology and geochemistry (2d ed.): New York, John Wiley and Sons, Inc., 477 p.

Chapelle, F.H., McMahon, P.B., Dubrovsky, N.M., Fuji, R.F., Oaksford, E.T., and Vroblesky, D.A., 1995, Deducing the distribution of terminal electron-accepting processes in hydrologically diverse groundwater systems: Water Resources Research, v. 31, no. 2, p. 359–371.

Clark, I.D., and Fritz, P., 1997, Environmental isotopes in hydrogeology: New York, Lewis Publishers, 328 p.

Cook, P.G., and Böhlke, J.K., 2000, Determining timescales for groundwater flow and solute transport, in Cook, P.G., and Herczeg, A., eds., Environmental tracers in subsurface hydrology: Boston, Kluwer Academic Publishers, p. 1–30.

Craig, H., and Lal, D., 1961, The production rate of natural tritium: Tellus, v. 13, p. 85–105.

Dasgupta, P.K., Dyke, J.V., Kirk, A.B., and Jackson, W.A., 2006, Perchlorate in the United States—Analysis of relative source contributions to the food chain: Environmental Science & Technology, v. 40, p. 6608–6614.

Figuers, S., 1998, Groundwater study and water supply history of the East Bay Plain, Alameda and Contra Costa Counties, CA: Livermore, California, Norfleet Consultants, 90 p.

Fontes, J.C., and Garnier, J.M., 1979, Determination of the initial ^{14}C activity of the total dissolved carbon—A review of the existing models and a new approach: Water Resources Research, v. 15, no. 2, p. 399–413.

Fram, M.S., and Belitz, Kenneth, 2011, Probability of detecting perchlorate under natural conditions in deep groundwater in California and the Southwestern United States: Environmental Science & Technology, v. 45, p. 1,271–1,277.

Fram, M.S., and Belitz, Kenneth, 2012, Status and understanding of groundwater quality in the Tahoe-Martis, Central Sierra, and Southern Sierra study units, 2006–2007—California GAMA Priority Basin Project: U.S. Geological Survey Scientific Investigations Report 2011-5216, 222 p.

Gilliom, R.J., Barbash, J.E., Crawford, C.G., Hamilton, P.A., Martin, J.D., Nakagaki, N., Nowell, L.H., Scott, J.C., Stackelberg, P.E., Thelin, G.P., and Wolock, D.M., 2006, The quality of our nation's waters—Pesticides in the nation's streams and ground water, 1992–2001: U.S. Geological Survey Circular 1291, 172 p.

Hanson, R.T., Li, Zhen, and Faunt, C.C., 2004, Documentation of the Santa Clara Valley regional ground-water/surface-water flow model, Santa Clara County, California: U.S. Geological Survey Scientific Investigations Report 2004-5231, 75 p.

Helsel, D.R., and Hirsch, R.M., 2002, Statistical methods in water resources: U.S. Geological Survey Techniques of Water-Resources Investigations, book 4, chap. A3, 510 p. (Also available at http://water.usgs.gov/pubs/twri/twri4a3/.)

Hem, J.D., 1985, Study and interpretation of the chemical characteristics of natural water (2d ed.): U.S. Geological Survey Water-Supply Paper 1473, 363 p.

Isaaks, E.H., and Srivastava, R.M., 1989, Applied geostatistics: New York, Oxford University Press, 511 p.

Ivahnenko, Tammy, and Barbash, J.E., 2004, Chloroform in the hydrologic system—Sources, transport, fate, occurrence, and effects on human health and aquatic organisms: U.S. Geological Survey Scientific Investigations Report 2004-5137, 34 p.

Jennings, C.W., 1977, Geologic map of California: California Department of Conservation, Division of Mines and Geology, Geologic Data Map No. 2, scale 1:750,000.

Johnson, T.D., and Belitz, Kenneth, 2009, Assigning land use to supply wells for the statistical characterization of regional groundwater quality—Correlating urban land use and VOC occurrence: Journal of Hydrology, v. 370, p. 100–108.

Jurgens, B.C., McMahon, P.B., Chapelle, F.H., and Eberts, S.M., 2009, An Excel® workbook for identifying redox processes in ground water: U.S. Geological Survey Open-File Report 2009-1004, 8 p. (Also available at http://pubs.usgs.gov/of/2009/1004/.)

Kulongoski, J.T., and Belitz, Kenneth, 2004, Ground-Water Ambient Monitoring and Assessment Program: U.S. Geological Survey Fact Sheet 2004-3088, 2 p. (Also available at http://pubs.usgs.gov/fs/2004/3088/.)

Kulongoski, J.T., Belitz, Kenneth, Landon, M.K., and Farrar, Christopher, 2010, Status and understanding of groundwater quality in the North San Francisco Bay Groundwater Basins, 2004—California GAMA Priority Basin Project: U.S. Geological Survey Scientific Investigations Report 2010-5089, 87 p. (Also available at http://pubs.usgs.gov/sir/2010/5089.)

Landon, M.K., Belitz, Kenneth, Jurgens, B.C., Kulongoski, J.T., and Johnson, T.D., 2010, Status and understanding of groundwater quality in the Central–Eastside San Joaquin Basin, 2006—California GAMA Priority Basin Project: U.S. Geological Survey Scientific Investigations Report 2009-5266, 97 p. (Also available at http://pubs.usgs.gov/sir/2009/5266/.)

Lucas, L.L., and Unterweger, M.P., 2000, Comprehensive review and critical evaluation of the half-life of tritium: Journal of Research of the National Institute of Standards and Technology, v. 105, no. 4, p. 541–549.

McMahon, P.B., and Chapelle, F.H., 2008, Redox processes and water quality of selected principal aquifer systems: Ground Water, v. 46, no. 2, p. 259–271.

Metzger, L.F., and Fio, J.L., 1997, Ground-water development and the effects on ground-water levels and water quality in the Town of Atherton, San Mateo County, California: U.S. Geological Survey Water-Resources Investigations Report 97-4033, 31 p. Available at http://pubs.usgs.gov/wri/1997/4033/report.pdf.

Michel, R.L., 1989, Tritium deposition in the continental United States, 1953–83: U.S. Geological Survey Water-Resources Investigations Report 89-4072, 46 p.

Michel, R.L., and Schroeder, R., 1994, Use of long-term tritium records from the Colorado River to determine timescales for hydrologic processes associated with irrigation in the Imperial Valley, California: Applied Geochemistry, v. 9, p. 387–401.

Moran, J.E., Hudson, G.B., Eaton, G.F., and Leif, R., 2002, A contamination vulnerability assessment for the Santa Clara and San Mateo groundwater basins: UCRL-TR-201929, 49 p.

Moran, M.J., Zogorski, J.S., and Squillace, P.J., 2007, Chlorinated solvents in groundwater of the United States: Environmental Science & Technology, v. 41, no. 1, p. 74–81.

Mueller, D.K., and Helsel, D.R., 1996, Nutrients in the Nations's waters—Too much of a good thing?: U.S. Geological Survey Circular 1136, 24 p.

Nakagaki, N., Price, C.V., Falcone, J.A., Hitt, K.J., and Ruddy, B.C., 2007, Enhanced National Land Cover Data 1992 (NLCDe 92): U.S. Geological Survey raster digital data, available online at http://water.usgs.gov/lookup/getspatial?nlcde92.

Nakagaki, N., and Wolock, D.M., 2005, Estimation of agricultural pesticide use in drainage basins using land cover maps and county pesticide data: U.S. Geological Survey Open-File Report 2005-1188, 56 p.

Piper, A.M., 1944, A graphic procedure in the geochemical interpretation of water analyses: American Geophysical Union Transactions, v. 25, p. 914–923.

Plummer, L.N., Michel, R.L., Thurman, E.M., and Glynn, P.D., 1993, Environmental tracers for age-dating young ground water, *in* Alley, W.M., ed., Regional groundwater quality: New York, Van Nostrand Reinhold, p. 255–294.

PRISM Climate Group, Oregon State University, 2010, United States average annual precipitation, maximum and minimum temperature, 1971–2009, accessed January 14, 2011, at http://prism.oregonstate.edu/.

Rajagopalan, S., Anderson, T.A., Cox, S., Harvey, G., Cheng, Q., and Jackson, W.A., 2009, Perchlorate in wet deposition across North America: Environmental Science & Technology, v. 43, p. 616–622.

Rajagopalan, S., Anderson, T.A., Fahlquist, L., Rainwater, K.A., Ridley, M., and Jackson, W.A., 2006, Widespread presence of naturally occurring perchlorate in the High Plains of Texas and New Mexico: Environmental Science & Technology, v. 40, p. 3156–3162.

Ray, M.C., Kulongoski, J.T., and Belitz, Kenneth, 2009, Ground-water quality data in the San Francisco Bay study unit, 2007—Results from the California GAMA Program: U.S. Geological Survey Data Series 396, 92 p.

Reimann, C., and de Caritat, P., 1998, Chemical elements in the environment—Factsheets for the Geochemist and Environmental Scientist: Berlin, Springer-Verlag, 398 p.

Rowe, B.L., Toccalino, P.L., Moran, M.J., Zogorski, J.S., and Price, C.V., 2007, Occurrence and potential human-health relevance of volatile organic compounds in drinking water from domestic wells in the United States: Environmental Health Perspectives, v. 115, no. 11, p. 1539–1546.

Santa Clara Valley Water District, 2011, Study of potential for groundwater contamination from past dry cleaner operations in Santa Clara County, 89 p., accessed February 1, 2012, at http://www.swrcb.ca.gov/sanfranciscobay/publications_forms/documents/SCVWD_Study/Study.pdf.

Saucedo, G.J., Bedford, D.R., Raines, G.L., Miller, R.J., and Wentworth, C.M., 2000, GIS data for the geologic map of California (version 2.0): Sacramento, California, California Department of Conservation, Division of Mines and Geology.

Scott, J.C., 1990, Computerized stratified random site selection approaches for design of a groundwater quality sampling network: U.S. Geological Survey Water-Resources Investigations Report 90-4101, 109 p.

Sparks, D.L., 1995, Environmental soil chemistry: San Diego, CA, Academic Press, 353 p.

State of California, 1999, Supplemental report of the 1999 Budget Act 1999–00 Fiscal Year, Item 3940-001-0001, State Water Resources Control Board, accessed August 2010 at http://swrcb.ca.gov/gama/docs/ab_599_bill_20011005_chaptered.pdf.

State of California, 2001a, Assembly Bill No. 599, Chapter 522, accessed August 2010 at http://www.lao.ca.gov/1999/99-0_supp_rpt_lang.html#3940.

State of California, 2001b, Groundwater Monitoring Act of 2001: California State Water Code, division 6, part 2.76, sections 10780–10782.3, accessed August 2010 at http://www.leginfo.ca.gov/cgibin/.

Sturchio, N.C., Böhlke, J.K., Beloso, A.D., Jr., Streger, S.H., Heraty, L.J., and Hatzinger, P.B., 2007, Oxygen and chlorine isotopic fractionation during perchlorate biodegradation—Laboratory results and implications for forensics and natural attenuation studies: Environmental Science & Technology, v. 41, no. 8, p. 2796–2802.

Toccalino, P.L., and Norman, J.E., 2006, Health-based screening levels to evaluate U.S. Geological Survey ground-water quality data: Risk Analysis, v. 26, no. 5, p. 1339–1348.

Toccalino, P.L., Norman, J.E., Phillips, R.H., Kauffman, L.J., Stackelberg, P.E., Nowell, L.H., Krietzman, S.J., and Post, G.B., 2004, Application of health-based screening levels to groundwater quality data in a state-scale pilot effort: U.S. Geological Survey Scientific Investigations Report 2004-5174, 64 p.

Torgersen, T., Clarke, W.B., and Jenkins, W.J., 1979, The tritium/helium-3 method in hydrology: IAEA-SM-228, v. 49, p. 917–930.

U.S. Department of Commerce, 2010, National Climatic Data Center, accessed January 11, 2010, at http://www.ncdc.noaa.gov/oa/ncdc.html.

U.S. Environmental Protection Agency, 1998, Code of Federal Regulations, title 40—Protection of environment, chapter 1—Environmental protection agency, subchapter E—Pesticide programs, part 159—Statements of policies and interpretations, subpart D—Reporting requirements for risk/benefit information, 40 CFR 159.184: National Archives and Records Administration, September 19, 1997; amended June 19, 1998, accessed September 5, 2008, at http://www.epa.gov/EPA-PEST/1997/September/Day-19/p24937.htm.

U.S. Environmental Protection Agency, 2005, List of known perchlorate releases in the U.S., March 25, 2005, accessed November 2007 at http://www.epa.gov/swerffrr/documents/perchlorate_releases_us_20050325.htm.

U.S. Environmental Protection Agency, 2006, 2006 Edition of the drinking water standards and health advisories: Washington, D.C., U.S. Environmental Protection Agency, Office of Water EPA/822/R-06–013. (Also available at http://www.epa.gov/waterscience/criteria/drinking/dwstandards.pdf.)

Vogel, J.C., and Ehhalt, D., 1963, The use of the carbon isotopes in groundwater studies: Radioisotopes in Hydrology, IAEA, p. 383–395.

Wallace, R.E., ed., 1990, The San Andreas Fault System, California: U.S. Geological Survey Professional Paper 1515, 304 p., accessed at http://pubs.usgs.gov/pp/1990/1515/pp1515.pdf.

Zogorski, J.S., Carter, J.M., Ivahnenko, Tamara, Lapham, W.W., Moran, M.J., Rowe, B.L., Squillace, P.J., and Toccalino, P.L., 2006, The quality of our Nation's waters—Volatile organic compounds in the Nation's ground water and drinking-water supply wells: U.S. Geological Survey Circular 1292, 101 p.

Appendix A. Use of Data from The California Department of Public Health (CDPH) Database

For the SFBAY study unit, the historical CDPH database contains more than 26,000 records of chemistry data for samples from more than 400 wells, requiring targeted retrievals to manageably use the data to assess water quality. The following paragraphs summarize the selection process for wells and data from the CDPH database for use in the grid-based status assessment.

The grid-based calculation of aquifer-scale proportion uses one value per grid cell. Where USGS data for inorganic constituents were not available, additional data to represent a cell were selected from the CDPH database. Of the 68 grid cells in the SFBAY study unit, 3 cells had USGS-grid wells with the full complement of inorganic constituent data collected by USGS-GAMA, 40 cells had USGS-grid wells with data for just nutrients, and 25 cells did not have USGS-grid wells. The CDPH database was queried to provide these missing data for inorganic constituents. CDPH wells with data for the most recent 3 years available at the time of sampling (April 1, 2004–March 31, 2007) were considered. If a well had more than one analysis for a constituent in the 3-year interval, then the most recent data were selected.

The data in the CDPH database are of unknown quality, and the database does not contain data for quality-control samples with which to make a comprehensive quality-control assessment of the data. Cation-anion imbalance was used as a rough quality-control metric. Because water is electrically neutral and must have a balance between positive (cations) and negative (anions) electrically charged dissolved species, the cation/anion imbalance commonly is used as a quality-assurance check for water-sample analysis (Hem, 1985). An imbalance of less than 10 percent was defined as indicating acceptable quality of the major ion data for the sample. It was assumed that if the quality of the major ion data were acceptable, then the quality of the data for the other inorganic constituents also would be acceptable. In practice, however, some wells did not have data for major ion constituents, so the cation-anion imbalance could not be evaluated.

The first choice was to select CDPH inorganic data for the grid well sampled by the USGS for other constituents, provided the CDPH data met the cation/anion balance criteria. This approach resulted in the selection of supplemental inorganic data from the CDPH database for 23 USGS-grid wells. To identify the USGS-grid wells that incorporated CDPH inorganic data, a well ID was created that added "DG" to the GAMA ID for these data (for example, SF-04 with CDPH data was assigned the well identification SF-DG-04; table A1).

If the first step did not yield CDPH inorganic data for the USGS-grid well, the second step was to search the CDPH database to identify the highest-ranked well with a cation/anion imbalance less than 10 percent in each grid cell. This step resulted in selecting CDPH inorganic data from additional CDPH wells that were not USGS-grid wells for 16 grid cells. To identify the data from these new CDPH-grid wells, an ID was created that added "DPH" after the study unit prefix. For cells that contained a USGS-grid well, the identification number of the CDPH-grid well remains the same as that of the USGS-grid well identifier (for example, CDPH-grid well SF-DPH-07; table A1). For cells that did not contain a USGS-grid well, the CDPH-grid well was given a sequential number starting after the last GAMA ID for the study unit (for example, CDPH-grid well SF-DPH-44; table A1).

If no wells in a grid cell met the cation/anion balance criteria or if there were insufficient data to evaluate charge balance, the third choice was to select the highest CDPH well that was randomly ranked with any of the needed inorganic data. This resulted in selecting CDPH inorganic data for seven additional wells. These new CDPH-grid wells were labeled using the same prefix as the other new CDPH-grid wells.

The result of these steps was that 47 of the 68 grid cells were represented by inorganic constituent data from the USGS database, the CDPH database, or both databases. In some cases, to achieve one value for each constituent per cell, it was necessary to select an additional well in a cell for certain constituents; hence, some cells have data from multiple CDPH wells. Inorganic data from the CDPH database were used for 34 grid cells (figs. A1A, A1B). CDPH data were available for 29 or 30 grid cells for most inorganic constituents, with the exception of data for nutrients, boron, lead, molybdenum, strontium, vanadium, and radioactive constituents, which were available for 0 to 25 wells (table 2). Estimates of aquifer-scale proportion for constituents based on a smaller number of wells are subject to a larger error associated with the 90-percent confidence intervals (on the basis of Jeffreys interval for the binomial distribution).

Differences in constituent laboratory reporting levels (LRLs) or method detection levels (MDLs) associated with USGS and CDPH data did not affect analysis of high or moderate relative-concentrations because concentrations greater than one-half of water-quality benchmarks were substantially higher than the reporting levels. Several types of comparisons between USGS-collected and CDPH data are described in Appendix E.

Table A1.   Cell number and USGS GAMA well identification numbers for well data used in the San Francisco Bay study unit, California GAMA Priority Basin Project.

[A USGS GAMA well identification number indicates the use of USGS data from the grid well; a CDPH GAMA well identification number with 'DG' indicates the use of CDPH inorganic data from the grid well; a CDPH GAMA well identification number with 'DPH' indicates the use of CDPH data from a different well. SF, San Francisco Bay study unit well; SFU or SFM, understanding well; –, no wells sampled or selected; na, not applicable]

USGS GAMA well identification number	Grid cell number	CDPH GAMA well identification number	Grid supplemented by CDPH data from different well	USGS GAMA well identification number	Grid cell number	CDPH GAMA well identification number	Grid supplemented by CDPH data from different well
Grid wells				**Grid wells**			
SF-01	1	–	–	None	29	–	SF-DPH-44
SF-02	2	–	–	None	32	–	SF-DPH-45
SF-03	15	–	–	None	35	–	SF-DPH-46
SF-04	16	SF-DG-04	–	None	68	–	SF-DPH-47
SF-05	17	SF-DG-05	–	**Understanding production wells**			
SF-06	18	–	–	SFU-01	17	–	–
SF-07	34	–	SF-DPH-07	SFU-02	48	–	–
SF-08	33	–	SF-DPH-08	SFU-03	48	–	–
SF-09	36	SF-DG-09	SF-DPH-09	SFU-04	49	–	–
SF-10	37	SF-DG-10	SF-DPH-10	SFU-05	49	–	–
SF-11	50	SF-DG-11	–	SFU-06	58	–	–
SF-12	51	SF-DG-12	–	SFU-07	58	–	–
SF-13	49	SF-DG-13	–	SFU-08	63	–	–
SF-14	48	SF-DG-14	SF-DPH-14	SFU-09	27	–	–
SF-15	47	SF-DG-15	–	SFU-10	27	–	–
SF-16	46	SF-DG-16	SF-DPH-16	SFU-11	na[1]	–	–
SF-17	45	SF-DG-17	SF-DPH-17	SFU-12	na[1]	–	–
SF-18	44	SF-DG-18	SF-DPH-18	**Monitor wells**			
SF-19	53	–	–	SFM-A1	1	–	–
SF-20	54	SF-DG-20	SF-DPH-20	SFM-A2	1	–	–
SF-21	55	SF-DG-21	–	SFM-A3	1	–	–
SF-22	56	SF-DG-22	SF-DPH-22	SFM-A4	1	–	–
SF-23	57	SF-DG-23	SF-DPH-23	SFM-B1	16	–	–
SF-24	58	–	–	SFM-B2	16	–	–
SF-25	59	SF-DG-25	SF-DPH-25	SFM-C1	58	–	–
SF-26	67	SF-DG-26	–	SFM-C2	58	–	–
SF-27	66	SF-DG-27	SF-DPH-27	SFM-C3	58	–	–
SF-28	64	–	–	SFM-C4	58	–	–
SF-29	65	SF-DG-29	–	SFM-C5	58	–	–
SF-30	60	SF-DG-30	–	SFM-D1	28	–	–
SF-31	63	SF-DG-31	–	SFM-D2	28	–	–
SF-32	62	SF-DG-32	SF-DPH-32	SFM-D3	28	–	–
SF-33	61	SF-DG-33	–	SFM-D4	28	–	–
SF-34	43	–	–	SFM-E1	28	–	–
SF-35	28	–	–	SFM-E2	28	–	–
SF-36	27	SF-DG-36	–	SFM-E3	28	–	–
SF-37	26	SF-DG-37	–	SFM-F1	23	–	–
SF-38	23	–	–	SFM-F2	23	–	–
SF-39	24	SF-DG-39	–	SFM-F3	23	–	–
SF-40	25	–	SF-DPH-40	SFM-F4	23	–	–
SF-41	11	–	–	SFM-F5	23	–	–
SF-42	10	–	–	SFM-F6	23	–	–
SF-43	13	–	–				

[1]These wells were outside the study unit boundaries.

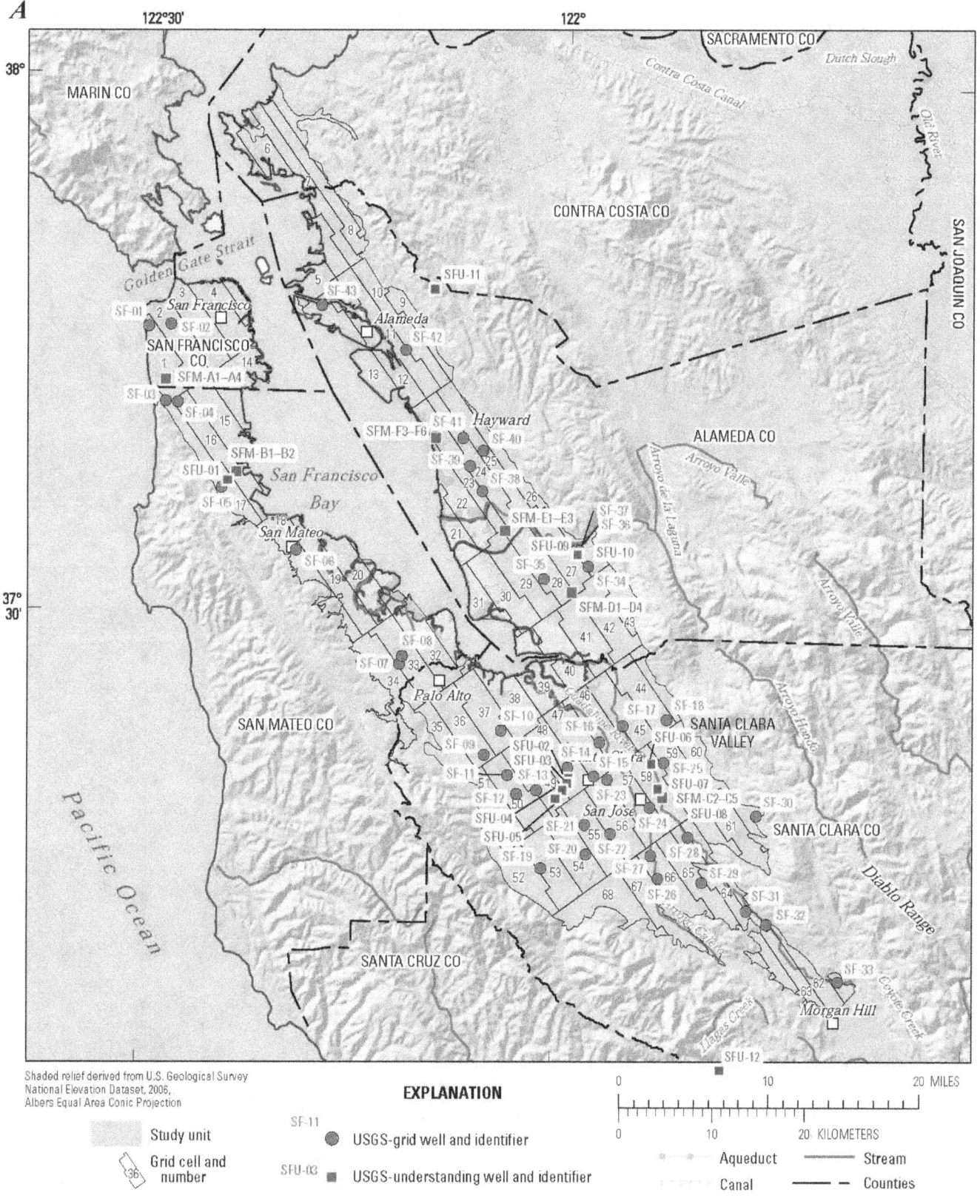

Figure A1. Identifiers and locations of (A) USGS-grid and USGS-understanding wells sampled during April to June 2007 and (B) CDPH-grid wells, San Francisco Bay study unit, California GAMA Priority Basin Project.

B

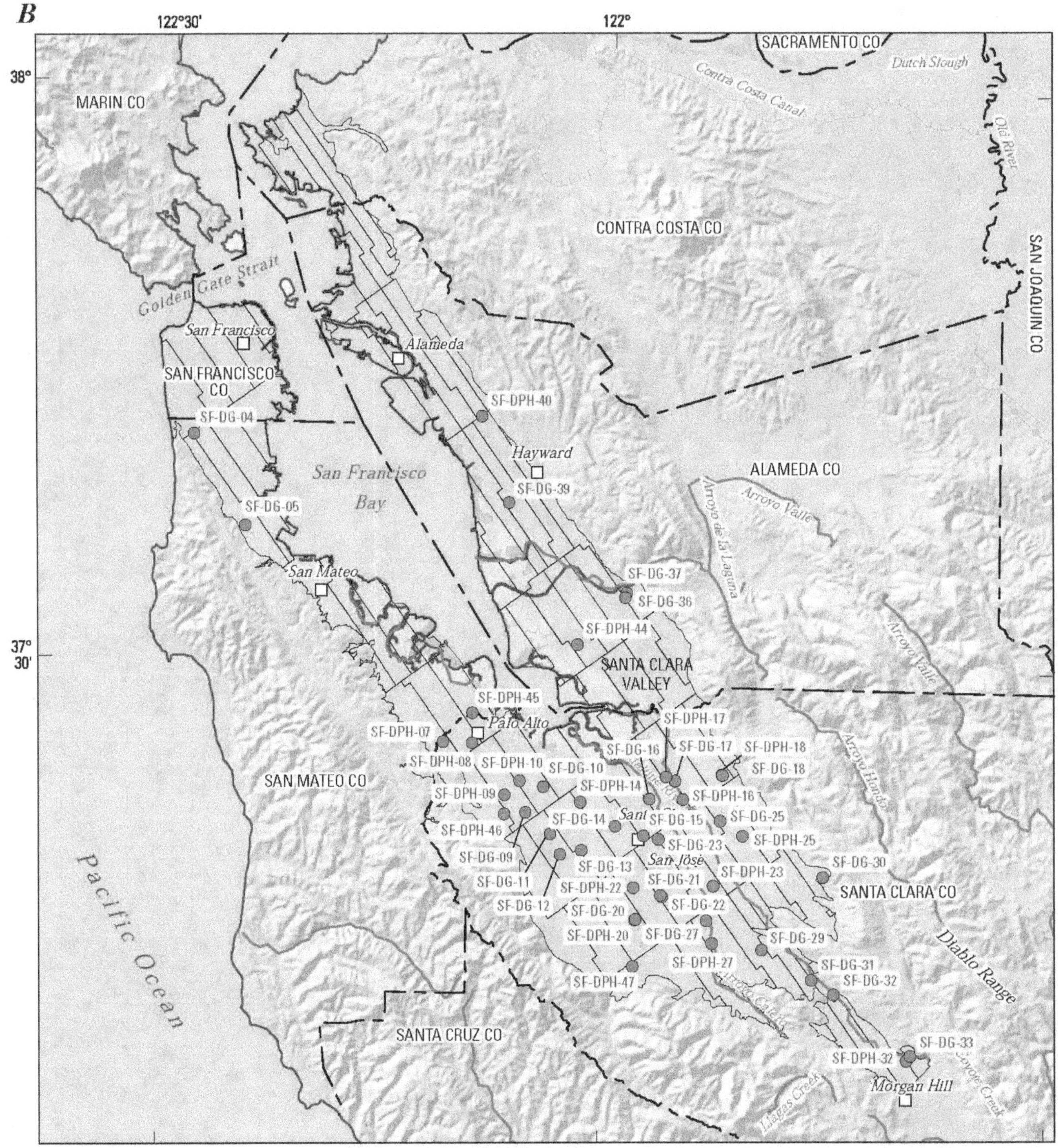

Shaded relief derived from U.S. Geological Survey National Elevation Dataset, 2006, Albers Equal Area Conic Projection

0 10 20 MILES

0 10 20 KILOMETERS

EXPLANATION

Study unit

SF-DG-27 CDPH-grid well and identifier

Aqueduct

Canal

Stream

Counties

Figure A1.—Continued

Appendix B. Calculating Total Dissolved Solids

Specific conductance, an electrical measure of total dissolved solids (TDS), was available for all 79 grid and understanding wells sampled by the USGS, whereas TDS was only measured for 30 of these wells. As a result, TDS values were calculated for wells that had no measured TDS from specific conductance (SC) values using a linear regression equation (TDS = 0.56*SC +58) developed from the SC and TDS relations from the 30 wells with TDS data so that all grid wells would have TDS values. The predicted TDS values using the regression equation closely matched measured TDS values ($R^2 = 0.99$). TDS values from CDPH were combined with USGS measured and calculated TDS values.

Appendix C. Estimation of Aquifer-Scale Proportions

Two statistical approaches, grid-based and spatially weighted, were selected to evaluate the proportions of the primary aquifer system in the SFBAY study unit that had high, moderate, or low relative-concentrations (compared to benchmarks) of constituents. The grid-based and spatially weighted estimations of aquifer-scale proportions, based on a spatially distributed grid cell network across the study unit, are intended to characterize the water quality of the primary aquifer system, or at depths from which drinking water is usually drawn. These approaches assign weights to wells based upon a single well per cell (grid-based) or the number of wells per cell (spatially weighted). Raw detection frequencies, derived from the percentage of the total number of wells with high or moderate relative-concentrations, also were calculated for individual constituents, but were not used for estimating aquifer-scale proportion because this method creates spatial bias towards regions with large numbers of wells.

1. Grid-based: One well in each grid cell, a "grid well," was randomly selected to represent the primary aquifer system (Belitz and others, 2010). For organic constituents, the one value in each grid cell was obtained from samples analyzed by USGS-GAMA from 43 grid wells. For inorganic constituents, the one value in each grid cell was obtained from samples analyzed by USGS-GAMA and data selected from the CDPH database. The relative-concentration for each constituent (concentration relative to its benchmark) was then evaluated for each grid well. The proportion of the primary aquifer system that had high relative-concentrations was calculated by dividing the number of cells with concentrations greater than the benchmark (relative-concentration greater than 1) by the number of grid wells with data in the SFBAY study unit. Proportions containing moderate and low relative-concentrations were calculated similarly. Confidence intervals for grid-based aquifer proportions were computed using the Jeffreys interval for the binomial distribution (Brown and others, 2001). The grid-based estimate is spatially unbiased. However, the grid-based approach may not identify constituents that exist at high concentrations in small proportions of the primary aquifer system.

The grid-based aquifer-scale proportions for constituent classes also are calculated on a one-value-per-cell basis. A cell with a high value for any constituent in the class is defined as a high cell, and the high proportion is the number of high cells divided by the number of cells with data for any of the constituents in the class. The moderate proportion for the constituent class is calculated similarly, except that a cell already defined as high cannot also be defined as moderate. A cell with a moderate value for any constituent in the class that does not also have a high value for any constituent in the class is defined as moderate.

2. Spatially weighted: The spatially weighted approach relied upon USGS-grid well data, CDPH data from April 1, 2004, to March 31, 2007 (most recent analyses per well for all wells within each grid cell), and selected USGS-understanding public-supply well data. However, instead of data from only one well per grid cell, the spatially weighted approach calculates the high, moderate, and low relative-concentrations for all of the wells in each cell. The high, moderate, and low aquifer-scale proportions are calculated for each cell, and then the proportions are averaged for all of the cells with data for the constituent (Isaaks and Srivastava, 1989). The resulting proportions are spatially unbiased. Confidence intervals for spatially weighted estimates of aquifer-scale proportion are not described in this report.

The raw detection frequency approach is merely the percentage (frequency) of wells within the study unit that had high relative-concentrations. It was calculated by considering all of the available data in the period from April 1, 2004, to March 31, 2007, for the USGS-grid well data, the CDPH-well data (the most recent analysis per well for all wells), and USGS-understanding wells. However, this approach could be spatially biased because the CDPH and USGS-understanding wells are not uniformly distributed. Consequently, high values (or low values) for wells clustered in a particular area represent a small part of the primary aquifer system, and could be given a disproportionately high (or low) weight compared to that given by spatially unbiased approaches. Raw detection frequencies of high relative-concentrations are provided to identify constituents for discussion in this report, but were not used to assess aquifer-scale proportions.

Appendix D. Ancillary Datasets

Land-use classifications and percentages of each classification, well construction information, geochemical conditions, and groundwater age data and classifications are listed in tables D1–D5.

Land-Use Classification

Land use was classified by using an enhanced version of the satellite-derived [98-ft (30-m) pixel resolution] USGS Enhanced National Land Cover Dataset (Nakagaki and others, 2007). This dataset has been used in previous national and regional studies relating land use to water quality (Gilliom and others, 2006; Zogorski and others, 2006). The dataset characterizes land cover during the early 1990s. The imagery was classified into 25 land-cover categories (Nakagaki and Wolock, 2005). These 25 land-cover categories were aggregated into three principal land-use classes for the purpose of general characterization of land use: urban, agricultural, and natural. Average land use (proportions of urban, agricultural, and natural) for the study unit, for each study area, and for areas within a radius of 1,640 ft (500 m) surrounding each well (table D1) were calculated using ArcGIS (version 9.2) (Johnson and Belitz, 2009).

Well Construction Information

Well construction data primarily were obtained from driller's logs filed with CDWR. Other sources of well construction data were ancillary records from well owners and the USGS National Water Information System database. Well construction data are not available in the CDPH database. Well identification verification procedures are described by Ray and others (2009). Well depths, depths to the tops and bottoms of the perforated intervals, and lengths of the perforated intervals for wells are listed in table D2.

Classification of Geochemical Condition

Geochemical conditions investigated as potential explanatory variables in this report include oxidation-reduction (redox) characteristics and pH (table D3).

Oxidation-reduction (redox) conditions influence the mobility of many organic and inorganic constituents (McMahon and Chapelle, 2008). Along groundwater flow paths, redox conditions commonly proceed along a well-documented sequence of terminal electron acceptor processes (TEAPs); one TEAP typically dominates at a particular time and aquifer location (Chapelle and others, 1995; Chapelle, 2001). The predominant TEAPs are oxygen-reduction, nitrate-reduction, manganese-reduction, iron-reduction, sulfate-reduction, and methanogenesis. The presence of redox-sensitive chemical species suggesting more than one TEAP may indicate mixed waters from different redox zones upgradient of the well, a well screened across more than one redox zone, or spatial heterogeneity in microbial activity in the aquifer.

Redox conditions were assigned to each sample using a modified version of the classification scheme of McMahon and Chapelle (2008) and Jurgens and others (2009) (tables D3, D4). Samples with DO > 0.5 mg/L were classified as oxic, and samples with DO ≤ 0.5 mg/L were classified as anoxic. The anoxic samples were further classified according to the TEAPs inferred from data for nitrate, manganese, and iron. Data for these constituents were obtained from USGS-GAMA where available and from the CDPH database ("DG" CDPH-grid wells). Inorganic constituent data were not available for all anoxic samples.

Groundwater Age Classification

Groundwater dating techniques provide a measure of the time since the groundwater was last in contact with the atmosphere (residence time in the aquifer). The techniques used in this report to estimate groundwater residence times or 'age' were those based on tritium (for example: Tolstikhin and Kamensky, 1969; Torgersen and others, 1979) and carbon-14 activities (for example: Vogel and Ehhalt, 1963; Plummer and others, 1993).

Tritium (^3H) is a short-lived radioactive isotope of hydrogen with a half-life of 12.32 years (Lucas and Unterweger, 2000). Tritium is produced naturally in the atmosphere from the interaction of cosmogenic radiation with nitrogen (Craig and Lal, 1961), by above-ground nuclear explosions, and by the operation of nuclear reactors. Tritium enters the hydrological cycle following oxidation to tritiated water. Natural background levels of tritium in precipitation are approximately 3 to 15 tritium units (TU) (Craig and Lal, 1961; Clark and Fritz, 1997). Above-ground nuclear explosions resulted in a large increase in tritium values in precipitation, beginning in about 1952 and peaking in 1963 at values over 1,000 TU in the northern hemisphere (Michel, 1989). Radioactive decay over a period of 60 years would decrease tritium values of 10 TU to 0.6 TU.

Previous investigations have used a range of tritium values from 0.3 to 1.0 TU as thresholds for indicating presence of water that has exchanged with the atmosphere since 1952 (Michel, 1989; Plummer and others, 1993; Michel and Schroeder, 1994; Clark and Fritz, 1997; Manning and others, 2005; Kulongoski and others, 2010; Landon and others, 2010). For samples collected for the SFBAY study unit in 2007, tritium values greater than a threshold of 1.0 TU were defined as indicating the presence of groundwater recharged since 1952 (Kulongoski and others, 2010; Landon and others, 2010). Water recharged since 1952 is defined as "modern" groundwater.

Carbon-14 (^{14}C) is a widely used chronometer based on the radiocarbon content of organic and inorganic carbon. Dissolved inorganic carbon species, carbonic acid, bicarbonate, and carbonate typically are used for ^{14}C dating of groundwater. ^{14}C is formed in the atmosphere by the interaction of cosmic-ray neutrons with nitrogen and, to a lesser degree, with oxygen and carbon, and by above-ground nuclear explosions. ^{14}C is incorporated into carbon dioxide and mixed throughout the atmosphere. The carbon dioxide enters the hydrologic cycle because it dissolves in precipitation and surface water in contact with the atmosphere. ^{14}C activity in groundwater, expressed as percent modern carbon (pmc), reflects the time since groundwater was last exposed to the atmospheric ^{14}C source. ^{14}C has a half-life of 5,730 years and can be used to estimate groundwater ages ranging from 1,000 to approximately 30,000 years before present.

The ^{14}C age (residence time, presented in years) is calculated on the basis of the decrease in ^{14}C activity as a result of radioactive decay since groundwater recharge, relative to an assumed initial ^{14}C concentration (Clark and Fritz, 1997). An average initial ^{14}C activity of 99 pmc is assumed for this study, with estimated errors on calculated groundwater ages up to ± 20 percent. Calculated ^{14}C ages in this study are referred to as "uncorrected" because they have not been adjusted to consider exchanges with sedimentary sources of carbon (Fontes and Garnier, 1979). Groundwater with a ^{14}C activity of >88 pmc is reported as having an age of <1,000 years; no attempt is made to refine ^{14}C ages <1,000 years. Measured values of pmc can be >100 in groundwater samples containing a significant component of water recharged after 1952 because the definition of pmc is based on ^{14}C activity in the absence of ^{14}C contributed by above-ground nuclear explosions (Clark and Fritz, 1997). For the SFBAY study unit, ^{14}C activity <90 pmc was defined as indicating the presence of groundwater recharged before the modern era (Kulongoski and others, 2010; Landon and others, 2010). Water recharged before the modern era is defined as "pre-modern" groundwater.

Tritium and ^{14}C data and age classifications are reported in table D5. Because of uncertainties in age distributions, particularly the uncertainties caused by mixing of waters of different ages in wells with long screened or open intervals and high withdrawal rates, the uncorrected ^{14}C ages were not specifically used for quantifying the relation between age and water quality in this report. While more sophisticated lumped parameter models for analyzing age distributions that incorporate mixing are available (Cook and Böhlke, 2000), use of these alternative models to understand age mixtures was not needed for the assessments in this report. Classification into modern, mixed, and pre-modern categories was sufficient to provide an appropriate and useful characterization for the purposes of examining groundwater quality.

For the SFBAY study unit, groundwater samples were assigned the following age classifications:

Classification	Tritium (TU)	^{14}C (pmc)
Modern	>1.0	>90
Modern or Mixed	>1.0	No data
Mixed	>1.0	<90
Mixed	<1.0	>90
Pre-modern	<1.0	<90

Table D1.   Well type, percent land-use by category, land-use classification, septic and leaking underground fuel tank (LUFT) density, and USGS-GAMA well identification number for GAMA well data and CDPH grid-well data used in the San Francisco Bay study unit, California GAMA Priority Basin Project.

[m, meter; tanks/km^2, tanks per square kilometer; na, not available; USGS, U.S. Geological Survey; DPH, California Department of Public Health; **Well types:** PSW, public supply well; IRR, irrigation well; MON, monitor well; INS, institutional well; IND, industrial well; DES, desalination well]

USGS-GAMA well identification number	Well type	Agricultural land use within 500 m of the well (percent)	Natural land use within 500 m of the well, percent	Urban land use within 500 m of the well, percent	Land-use classification	Septic density[1] (tanks/km^2)	LUFT density[2] (tanks/ km^2)
USGS-grid wells							
SF-01	IRR	0	61	39	Natural	0	1
SF-02	IRR	0	73	27	Natural	0	2
SF-03	PSW	0	2	98	Urban	0	1
SF-04	PSW	0	5	95	Urban	1	5
SF-05	PSW	0	3	97	Urban	0	2
SF-06	IRR	0	3	97	Urban	1	3
SF-07	INS	0	8	92	Urban	0	2
SF-08	IRR	0	5	95	Urban	0	2
SF-09	PSW	0	0	100	Urban	16	1
SF-10	PSW	0	3	97	Urban	22	5
SF-11	PSW	0	1	99	Urban	6	2
SF-12	PSW	0	2	98	Urban	8	1
SF-13	PSW	0	5	95	Urban	6	1
SF-14	PSW	0	2	98	Urban	5	1
SF-15	PSW	0	4	96	Urban	0	5
SF-16	PSW	0	2	98	Urban	0	12
SF-17	PSW	1	37	62	Urban	1	2
SF-18	PSW	0	0	100	Urban	0	0
SF-19	IRR	0	14	86	Urban	16	1
SF-20	PSW	0	1	99	Urban	34	2
SF-21	PSW	0	1	99	Urban	3	1
SF-22	PSW	0	2	98	Urban	10	3
SF-23	PSW	0	5	95	Urban	0	12
SF-24	PSW	0	2	98	Urban	3	6
SF-25	PSW	0	10	90	Urban	5	3
SF-26	PSW	7	19	74	Urban	0	1
SF-27	PSW	2	14	84	Urban	9	2
SF-28	PSW	0	34	66	Urban	1	1
SF-29	PSW	0	2	98	Urban	1	1
SF-30	PSW	17	67	16	Natural	5	0
SF-31	PSW	20	21	59	Urban	0	0
SF-32	PSW	1	81	18	Natural	3	0
SF-33	PSW	46	49	5	Natural	11	0
SF-34	IRR	0	57	43	Natural	0	1
SF-35	DES	0	15	85	Urban	0	2
SF-36	PSW	0	28	72	Urban	10	2
SF-37	PSW	0	29	71	Urban	14	3
SF-38	PSW	3	21	77	Urban	0	1
SF-39	PSW	0	15	85	Urban	69	2
SF-40	PSW	0	6	94	Urban	2	3
SF-41	IRR	0	14	86	Urban	5	6
SF-42	IND	0	5	95	Urban	0	13
SF-43	IRR	0	36	64	Urban	0	4

Table D1. Well type, percent land-use by category, land-use classification, septic and leaking underground fuel tank (LUFT) density, and USGS-GAMA well identification number for GAMA well data and CDPH grid-well data used in the San Francisco Bay study unit, California GAMA Priority Basin Project.—Continued

[m, meter; tanks/km²; tanks per square kilometer; na, not available; USGS, U.S. Geological Survey; DPH, California Department of Public Health; **Well types**: PSW, public supply well; IRR, irrigation well; MON, monitor well; INS, institutional well; IND, industrial well; DES, desalination well]

USGS-GAMA well identification number	Well type	Agricultural land use within 500 m of the well (percent)	Natural land use within 500 m of the well, percent	Urban land use within 500 m of the well, percent	Land-use classification	Septic density[1] (tanks/km²)	LUFT density[2] (tanks/km²)
CDPH-grid wells							
SF-DPH-44	PSW	0	5	95	Urban	na	na
SF-DPH-45	PSW	0	1	99	Urban	na	na
SF-DPH-46	PSW	0	3	97	Urban	na	na
SF-DPH-47	PSW	0	28	72	Urban	na	na
SF-DPH-48	PSW	0	0	100	Urban	na	na
SF-DPH-49	PSW	0	1	99	Urban	na	na
SF-DPH-50	PSW	0	1	99	Urban	na	na
SF-DPH-51	PSW	0	0	100	Urban	na	na
SF-DPH-52	PSW	69	7	24	Agricultural	na	na
SF-DPH-53	PSW	0	13	87	Urban	na	na
SF-DPH-54	PSW	0	2	98	Urban	na	na
SF-DPH-55	PSW	0	1	99	Urban	na	na
SF-DPH-56	PSW	0	1	99	Urban	na	na
SF-DPH-57	PSW	0	3	97	Urban	na	na
SF-DPH-58	PSW	92	6	2	Agricultural	na	na
SF-DPH-59	PSW	1	11	88	Urban	na	na
SF-DPH-60	PSW	0	25	75	Urban	na	na
SF-DPH-61	PSW	0	3	97	Urban	na	na
SF-DPH-62	PSW	0	22	78	Urban	na	na
USGS-understanding wells							
SFM-A1	MON	0	45	55	Urban	0	1
SFM-A2	MON	0	45	55	Urban	0	1
SFM-A3	MON	0	45	55	Urban	0	1
SFM-A4	MON	0	45	55	Urban	0	1
SFM-B1	MON	0	8	92	Urban	0	2
SFM-B2	MON	0	8	92	Urban	0	2
SFM-C1	MON	0	14	86	Urban	3	2
SFM-C2	MON	0	14	86	Urban	3	2
SFM-C3	MON	0	14	86	Urban	3	2
SFM-C4	MON	0	14	86	Urban	3	2
SFM-C5	MON	0	14	86	Urban	3	2
SFM-D1	MON	0	12	88	Urban	1	2
SFM-D2	MON	0	12	88	Urban	1	2
SFM-D3	MON	0	12	88	Urban	1	2
SFM-D4	MON	0	12	88	Urban	1	2
SFM-E1	MON	0	0	100	Urban	0	1
SFM-E2	MON	0	0	100	Urban	0	1
SFM-E3	MON	0	0	100	Urban	0	1
SFM-F1	MON	0	60	40	Natural	0	3
SFM-F2	MON	0	60	40	Natural	0	3
SFM-F3	MON	0	60	40	Natural	0	3
SFM-F4	MON	0	60	40	Natural	0	3
SFM-F5	MON	0	60	40	Natural	0	3
SFM-F6	MON	0	60	40	Natural	0	3
SFU-01	PSW	0	7	93	Urban	1	7
SFU-02	PSW	0	0	100	Urban	7	3
SFU-03	PSW	0	1	99	Urban	4	3

Table D1.   Well type, percent land-use by category, land-use classification, septic and leaking underground fuel tank (LUFT) density, and USGS-GAMA well identification number for GAMA well data and CDPH grid-well data used in the San Francisco Bay study unit, California GAMA Priority Basin Project.—Continued

[m, meter; tanks/km^2; tanks per square kilometer; na, not available; USGS, U.S. Geological Survey; DPH, California Department of Public Health; **Well types:** PSW, public supply well; IRR, irrigation well; MON, monitor well; INS, institutional well; IND, industrial well; DES, desalination well]

USGS-GAMA well identification number	Well type	Agricultural land use within 500 m of the well (percent)	Natural land use within 500 m of the well, percent	Urban land use within 500 m of the well, percent	Land-use classification	Septic density[1] (tanks/km^2)	LUFT density[2] (tanks/ km^2)
USGS-understanding wells							
SFU-04	PSW	0	2	98	Urban	0	1
SFU-05	PSW	0	12	88	Urban	0	1
SFU-06	PSW	0	12	88	Urban	2	4
SFU-07	PSW	0	10	90	Urban	0	2
SFU-08	PSW	0	23	77	Urban	1	7
SFU-09	PSW	0	29	71	Urban	9	2
SFU-10	PSW	0	24	76	Urban	8	2
SFU-11	PSW	0	98	2	Natural	3	0
SFU-12	PSW	0	99	1	Natural	3	0

[1] Septic tank density within 500-meter radius of well site, based on 1990 U.S. Census data.

[2] LUFT density within 500-meter radius of well site, based on GEIMS LUFT database (California Environmental Protection Agency, 2001).

Table D2.   Well construction information for wells used in the San Francisco Bay study unit, California GAMA Priority Basin Project.

[ft, feet; LSD, land-surface datum; USGS, U.S. Geological Survey; SF, San Francisco Bay study unit; DPH, California Department of Public Health; na, not available; **Well types**: PSW, public supply well; IRR, irrigation well; MON, monitor well; INS, institutional well; IND, industrial well; DES, desalination well]

| USGS-GAMA well identification number indicating data source | Well type | Construction information | | | | Aridity index[1] (dimensionless) |
		Well depth (ft below LSD)	Top of perforations (ft below LSD)	Bottom of perforations (ft below LSD)	Length from top of uppermost perforated interval to bottom of well (ft)	
USGS-grid wells						
SF-01	IRR	na	na	na	na	0.48
SF-02	IRR	360	170	350	190	0.54
SF-03	PSW	410	170	375	240	0.51
SF-04	PSW	na	na	na	na	0.54
SF-05	PSW	480	na	na	na	0.50
SF-06	IRR	180	na	na	na	0.45
SF-07	INS	220	na	na	na	0.37
SF-08	IRR	275	240	275	35	0.37
SF-09	PSW	1,120	289	1120	831	0.37
SF-10	PSW	680	290	660	390	0.31
SF-11	PSW	596	348	526	248	0.36
SF-12	PSW	na	na	na	na	0.37
SF-13	PSW	760	340	750	420	0.35
SF-14	PSW	528	165	363	363	0.32
SF-15	PSW	810	540	790	270	0.31
SF-16	PSW	665	295	665	370	0.31
SF-17	PSW	na	na	na	na	0.32
SF-18	PSW	816	300	816	516	0.34
SF-19	IRR	540	200	520	340	0.46
SF-20	PSW	840	358	798	482	0.38
SF-21	PSW	815	350	795	465	0.35
SF-22	PSW	827	378	818	449	0.35
SF-23	PSW	890	300	870	590	0.31
SF-24	PSW	780	265	774	515	0.32
SF-25	PSW	612	267	603	345	0.32
SF-26	PSW	427	107	376	320	0.37
SF-27	PSW	437	186	400	251	0.35
SF-28	PSW	517	385	454	132	0.33
SF-29	PSW	275	102	266	173	0.37
SF-30	PSW	80	41	80	39	0.47
SF-31	PSW	286	na	na	na	0.41
SF-32	PSW	na	na	na	na	0.42
SF-33	PSW	366	161	346	205	0.43
SF-34	IRR	153	60	153	93	0.35
SF-35	DES	248	216	240	32	0.34
SF-36	PSW	190	100	180	90	0.37
SF-37	PSW	200	80	177	120	0.37
SF-38	PSW	535	na	na	na	0.37
SF-39	PSW	600	480	580	120	0.40
SF-40	PSW	550	245	530	305	0.43
SF-41	IRR	155	35	155	120	0.42
SF-42	IND	495	324	479	171	0.45
SF-43	IRR	353	262	300	91	0.44

Table D2.   Well construction information for wells used in the San Francisco Bay study unit, California GAMA Priority Basin Project.—
Continued

[ft, feet; LSD, land-surface datum; USGS, U.S. Geological Survey; SF, San Francisco Bay study unit; DPH, California Department of Public Health; na, not available; **Well types**: PSW, public supply well; IRR, irrigation well; MON, monitor well; INS, institutional well; IND, industrial well; DES, desalination well]

USGS-GAMA well identification number indicating data source	Well type	Construction information				Aridity index[1] (dimensionless)
		Well depth (ft below LSD)	Top of perforations (ft below LSD)	Bottom of perforations (ft below LSD)	Length from top of uppermost perforated interval to bottom of well (ft)	
CDPH-grid wells						
SF-DPH-44	PSW	na	na	na	na	na
SF-DPH-45	PSW	608	181	532	427	na
SF-DPH-46	PSW	828	144	624	684	na
SF-DPH-47	PSW	320	142	301	178	na
SF-DPH-48	PSW	515	260	500	255	na
SF-DPH-49	PSW	na	na	na	na	na
SF-DPH-50	PSW	na	na	na	na	na
SF-DPH-51	PSW	na	299	823	na	na
SF-DPH-52	PSW	na	na	na	na	na
SF-DPH-53	PSW	na	260	716	na	na
SF-DPH-54	PSW	na	na	na	na	na
SF-DPH-55	PSW	na	358	790	na	na
SF-DPH-56	PSW	na	na	na	na	na
SF-DPH-57	PSW	na	440	531	na	na
SF-DPH-58	PSW	na	na	na	na	na
SF-DPH-59	PSW	na	na	na	na	na
SF-DPH-60	PSW	na	na	na	na	na
SF-DPH-61	PSW	na	na	na	na	na
SF-DPH-62	PSW	na	na	na	na	na
USGS-understanding wells						
SFM-A1	MON	575	555	565	20	0.49
SFM-A2	MON	440	410	430	30	0.49
SFM-A3	MON	270	240	260	30	0.49
SFM-A4	MON	155	140	150	15	0.49
SFM-B1	MON	146	126	136	20	0.45
SFM-B2	MON	74	54	64	20	0.45
SFM-C1	MON	1,000	820	840	180	0.32
SFM-C2	MON	640	620	640	20	0.32
SFM-C3	MON	540	520	540	20	0.32
SFM-C4	MON	425	405	425	20	0.32
SFM-C5	MON	72	62	72	10	0.32
SFM-D1	MON	480	450	480	30	0.34
SFM-D2	MON	340	330	340	10	0.34
SFM-D3	MON	260	230	260	30	0.34
SFM-D4	MON	80	50	80	30	0.34
SFM-E1	MON	470	430	470	40	0.35
SFM-E2	MON	200	180	200	20	0.35
SFM-E3	MON	100	50	100	50	0.35
SFM-F1	MON	1,010	990	1,010	20	0.42
SFM-F2	MON	860	830	860	30	0.42
SFM-F3	MON	640	530	640	110	0.42
SFM-F4	MON	318	298	318	20	0.42
SFM-F5	MON	138	128	138	10	0.42
SFM-F6	MON	45	35	45	10	0.42
SFU-01	PSW	630	na	na	na	0.46
SFU-02	PSW	570	309	557	261	0.32

Table D2. Well construction information for wells used in the San Francisco Bay study unit, California GAMA Priority Basin Project.—Continued

[ft, feet; LSD, land-surface datum; USGS, U.S. Geological Survey; SF, San Francisco Bay study unit; DPH, California Department of Public Health; na, not available; **Well types**: PSW, public supply well; IRR, irrigation well; MON, monitor well; INS, institutional well; IND, industrial well; DES, desalination well]

USGS-GAMA well identification number indicating data source	Well type	Construction information				Aridity index[1] (dimensionless)
		Well depth (ft below LSD)	Top of perforations (ft below LSD)	Bottom of perforations (ft below LSD)	Length from top of uppermost perforated interval to bottom of well (ft)	
USGS-understanding wells						
SFU-03	PSW	594	310	563	284	0.32
SFU-04	PSW	800	445	780	355	0.33
SFU-05	PSW	604	302	507	302	0.34
SFU-06	PSW	800	315	745	485	0.32
SFU-07	PSW	560	295	467	265	0.32
SFU-08	PSW	749	314	737	435	0.34
SFU-09	PSW	320	220	300	100	0.37
SFU-10	PSW	465	189	455	276	0.35
SFU-11	PSW	na	na	na	na	0.58
SFU-12	PSW	na	na	na	na	0.88

[1]Aridity index is average annual precipitation divided by average annual evapotranspiration.

Table D3.   pH and oxidation-reduction constituent concentrations and classifications for samples from the San Francisco Bay study unit, California GAMA Priority Basin Project.

[redox, oxidation-reduction; mg/L, milligram per liter; µg/L, microgram per liter; <, less than; >, greater than; ≥, greater than or equal to; >, greater than; <, less than; na, not available; O_2, oxygen; NO_3, nitrate; Mn, manganese; Fe, iron. Redox classifications: oxic, dissolved oxygen ≥ 0.5 mg/L; anoxic/suboxic, dissolved oxygen < 0.5 mg/L; NO_3-red, nitrate reducing conditions; Mn-red, manganese reducing conditions; MnFe-red, manganese and iron reducing conditions]

Well No.	Source of inorganic data[1]	pH (standard units)	Oxidation-reduction constituents					Redox classification
			Dissolved oxygen	Nitrate plus nitrite, as nitrogen	Magnanese	Iron	Sulfate	
Redox threshold value			≥0.5	>0.5	>50	>100	>4	
Possible redox type if concentration > redox threshold value			O_2	NO_3	Mn	Fe		
Analysis reporting level (unless otherwise noted)			0.2	0.06	0.18	5.0	0.18	
			mg/L	mg/L	µg/L	µg/L	mg/L	
Grid wells								
SF-01	CDPH	7.5	9.4	7.47	na	na	na	Oxic
SF-02	CDPH	7.3	6.3	12.7	na	na	na	Oxic
SF-03	CDPH	7.5	2.6	5.66	na	na	na	Oxic
SF-04	CDPH	8.1	1.7	2.02	41.0	<100	10.0	Oxic
SF-05	CDPH	7.1	4.8	1.29	na	na	29.0	Oxic
SF-06	CDPH	6.5	4.7	5.28	na	na	na	Oxic
SF-07	CDPH	6.9	1.9	6.58	na	na	na	Oxic
SF-08	CDPH	7.6	6.5	0.3	na	na	na	Oxic
SF-09	CDPH	7.3	4.8	7.66	na	na	41.1	Oxic
SF-10	CDPH	7.6	1.1	3.8	na	na	37.1	Oxic
SF-11	CDPH	7.3	11.9	2.96	<20	<100	32	Oxic
SF-12	CDPH	7.3	7.9	2.15	<20	<100	39	Oxic
SF-13	CDPH	7.1	5.8	6.2	<20	<100	21	Oxic
SF-14	CDPH	7.4	4.1	4.44	<20	<100	39	Oxic
SF-15	CDPH	7.7	<0.2	1.36	<20	<100	45	Anoxic (NO_3-red)
SF-16	CDPH	7.9	<0.2	0.52	27	<100	50	Anoxic (NO_3-red)
SF-17	CDPH	7.4	0.6	3.51	na	na	na	Oxic
SF-18	CDPH	7.4	5.1	5.33	88.9	101	72.7	Mixed
SF-19	CDPH	7.7	0.2	0.2	na	na	na	Anoxic
SF-20	CDPH	7.3	5.9	3.06	15.2	287	46.2	Mixed
SF-21	CDPH	7.2	4.9	4.85	5.08	173	35.9	Mixed
SF-22	CDPH	7.5	3.1	0.69	1.25	0	44.9	Oxic
SF-23	CDPH	7.6	3.8	1.93	<20	<100	60	Oxic
SF-24	USGS	7.4	1.1	3.81	0.70	11	64.9	Oxic
SF-25	CDPH	7.4	7.3	6.14	19.8	69.2	62.4	Oxic
SF-26	CDPH	7.3	0.8	0.56	3.76	73.5	32.0	Oxic
SF-27	CDPH	7.4	1.1	0.98	na	na	37.0	Oxic
SF-28	USGS	7.5	1.1	3.99	<0.2	<6	81.3	Oxic
SF-29	CDPH	7.4	1.8	2.91	na	na	68.0	Oxic
SF-30	CDPH	7.1	5.7	5.6	na	na	210	Oxic
SF-31	CDPH	7.5	2.1	1.2	na	na	37.0	Oxic
SF-32	CDPH	7.7	1.1	0.89	na	na	23.0	Oxic
SF-33	CDPH	7.3	3.3	3.43	na	na	32.0	Oxic
SF-34	CDPH	7.3	0.2	2.67	na	na	na	Anoxic
SF-35	USGS	7.2	0.5	0.49	408	24	142	Anoxic (Mn-red)

Table D3.   pH and oxidation-reduction constituent concentrations and classifications for samples from the San Francisco Bay study unit, California GAMA Priority Basin Project.—Continued

[redox, oxidation-reduction; mg/L, milligram per liter; µg/L, microgram per liter; <, less than; >, greater than; ≥, greater than or equal to; >, greater than; <, less than; na, not available; O_2, oxygen; NO_3, nitrate; Mn, manganese; Fe, iron. Redox classifications: oxic, dissolved oxygen ≥ 0.5 mg/L; anoxic/suboxic, dissolved oxygen < 0.5 mg/L; NO_3-red, nitrate reducing conditions; Mn-red, manganese reducing conditions; MnFe-red, manganese and iron reducing conditions]

			Oxidation-reduction constituents					
			Dissolved oxygen	Nitrate plus nitrite, as nitrogen	Magnanese	Iron	Sulfate	
Redox threshold value			≥0.5	>0.5	>50	>100	>4	
Possible redox type if concentration > redox threshold value			O_2	NO_3	Mn	Fe		
Analysis reporting level (unless otherwise noted)			0.2	0.06	0.18	5.0	0.18	
Well No.	Source of inorganic data[1]	pH (standard units)	mg/L	mg/L	mg/L	mg/L	mg/L	Redox classification
Grid wells—Continued								
SF-36	CDPH	7.1	2.7	1.27	na	na	72.0	Oxic
SF-37	CDPH	7.2	0.3	0.69	na	na	90.0	Anoxic
SF-38	CDPH	7.5	<0.2	<0.06	na	na	na	Anoxic
SF-39	CDPH	7.3	<0.2	0.15	92.0	<100	51.0	Anoxic (Mn-red)
SF-40	CDPH	7.4	0.4	2.64	na	na	na	Anoxic
SF-41	CDPH	9.4	10.5	E0.05	na	na	na	Oxic
SF-42	CDPH	7.2	0.7	2.04	na	na	na	Oxic
SF-43	CDPH	7.3	1.0	0.19	na	na	na	Oxic
Understanding wells								
SFM-A1	USGS	7.6	0.2	<0.06	239	202	81.5	Anoxic (MnFe-red)
SFM-A2	USGS	8.0	0.2	4.82	63.6	6	7.97	Anoxic (Mn-red)
SFM-A3	USGS	7.4	0.6	8.54	590	56	34.1	Mixed oxic/Mn-red
SFM-A4	USGS	7.5	3.0	11.2	0.60	E4	43.4	Oxic
SFM-B1	USGS	7.7	1.2	<0.06	264	106	33.6	Mixed oxic/MnFe-red
SFM-B2	USGS	6.9	0.2	<0.06	8,560	1,370	717	Anoxic (MnFe-red)
SFM-C1	USGS	8.0	0.2	<0.06	22.5	E4	18.0	Anoxic (suboxic)
SFM-C2	USGS	7.9	0.3	6.57	21.3	E4	58.0	Anoxic (NO_3-red)
SFM-C3	USGS	7.5	0.2	3.73	8.70	7	55.0	Anoxic (NO_3-red)
SFM-C4	USGS	7.6	0.3	3.85	10.4	10	62.5	Anoxic (NO_3-red)
SFM-C5	USGS	7.3	0.6	2.56	0.60	E6	87.4	Oxic
SFM-D1	USGS	8.0	0.2	<0.06	165	114	35.0	Anoxic (MnFe-red)
SFM-D2	USGS	7.7	0.2	<0.06	128	9	45.3	Anoxic (Mn-red)
SFM-D3	USGS	7.2	<0.2	<0.06	2,030	E12	114	Anoxic (Mn-red)
SFM-D4	USGS	7.2	<0.2	3.63	3.40	14	101	Anoxic (NO_3-red)
SFM-E1	USGS	7.8	0.2	0.28	99.9	E4	60.6	Anoxic (suboxic)
SFM-E2	USGS	7.4	0.3	11	29.6	E4	73.7	Anoxic (NO_3-red)
SFM-E3	USGS	7.2	0.7	3.56	8.9	19	129	Oxic
SFM-F1	USGS	7.8	5.5	<0.06	92.4	10	61.5	Mixed oxic/Mn-red
SFM-F2	USGS	8.2	0.1	<0.06	25.4	E5	37.2	Anoxic (suboxic)
SFM-F3	USGS	7.6	<0.2	0.36	55.1	<6	33.1	Anoxic (Mn-red)
SFM-F4	USGS	7.7	0.2	<0.06	152	E5	50.7	Anoxic (Mn-red)
SFM-F5	USGS	7.8	0.2	<0.06	90.3	<6	37.5	Anoxic (Mn-red)
SFM-F6	USGS	6.5	0.5	<0.06	37,000	<300	5,540	Mixed oxic/Mn-red
SFU-01	USGS	7.5	0.2	0.33	84.8	16	55.3	Anoxic (Mn-red)
SFU-02	CDPH	7.5	3.0	3.78	na	na	na	Oxic

Table D3.   pH and oxidation-reduction constituent concentrations and classifications for samples from the San Francisco Bay study unit, California GAMA Priority Basin Project.—Continued

[redox, oxidation-reduction; mg/L, milligram per liter; μg/L, microgram per liter; <, less than; >, greater than; ≥, greater than or equal to; >, greater than; <, less than; na, not available; O_2, oxygen; NO_3, nitrate; Mn, manganese; Fe, iron. Redox classifications: oxic, dissolved oxygen ≥ 0.5 mg/L; anoxic/suboxic, dissolved oxygen < 0.5 mg/L; NO_3-red, nitrate reducing conditions; Mn-red, manganese reducing conditions; MnFe-red, manganese and iron reducing conditions]

			Oxidation-reduction constituents					
			Dissolved oxygen	Nitrate plus nitrite, as nitrogen	Magnanese	Iron	Sulfate	
Redox threshold value			≥0.5	>0.5	>50	>100	>4	
Possible redox type if concentration > redox threshold value			O_2	NO_3	Mn	Fe		
Analysis reporting level (unless otherwise noted)			0.2	0.06	0.18	5.0	0.18	
Well No.	Source of inorganic data[1]	pH (standard units)	mg/L	mg/L	mg/L	mg/L	mg/L	Redox classification
Understanding wells—Continued								
SFU-03	CDPH	7.5	6.3	4.07	na	na	na	Oxic
SFU-04	CDPH	7.8	7.9	5.34	na	na	na	Oxic
SFU-05	CDPH	7.3	8.8	5.89	na	na	na	Oxic
SFU-06	USGS	7.5	0.4	1.14	<0.2	<6	72.9	Anoxic (NO_3-red)
SFU-07	USGS	7.4	2.4	3.3	<0.2	<6	77.1	Oxic
SFU-08	USGS	7.5	1.0	3.93	<0.2	7	78.0	Oxic
SFU-09	CDPH	7.2	1.9	2.73	na	na	na	Oxic
SFU-10	CDPH	7.4	0.2	2.02	na	na	na	Anoxic (NO_3-red)
SFU-11	CDPH	7.0	2.5	E0.06	na	na	na	Oxic
SFU-12	CDPH	6.8	5.0	0.11	na	na	na	Oxic

[1] Only data for manganese, iron, and/or sulfate were supplemented. Data for pH, dissolved oxygen, and nitrate were collected by the USGS at all wells.

Table D4.   Oxidation-reduction classification system applied to samples from the San Francisco Bay study unit, California GAMA Priority Basin Project.

[Oxidation-reduction classes: NO_3-red, nitrate-reducing; Mn-red, manganese-reducing; MnFe-red, manganese and iron reducing; Fe-red, iron-reducing. Other abbreviations: mg/L, milligram per liter; μg/L, microgram per liter; >, greater than; <, less than; na, data not available; any, any concentration]

Oxidation-reduction class	Dissolved oxygen (mg/L)	Nitrate, as nitrogen (mg/L)	Manganese (μg/L)	Iron (μg/L)
Oxic classifications				
Oxic	> 0.5	any	< 50	< 100
Oxic	> 0.5	any	na	na
Oxic	> 0.5	na	na	na
Anoxic classifications				
Anoxic (suboxic)	< 0.5	< 0.5	< 50	< 100
Anoxic (NO_3-red)	< 0.5	> 0.5	< 50	< 100
Anoxic (Mn-red)	< 0.5	< 0.5	> 50	< 100
Anoxic (MnFe-red)	< 0.5	< 0.5	> 50	> 100
Anoxic (Fe-red)	< 0.5	< 0.5	< 50	> 100
Anoxic	< 0.5	< 0.5	na	na
Anoxic (NO_3-red)	< 0.5	> 0.5	na	na
Anoxic	< 0.5	na	na	na
Mixed classifications				
Mixed (oxic/Mn-red)	> 0.5	any	> 50	< 100
Mixed (oxic/MnFe-red)	> 0.5	any	> 50	> 100

Table D5. Tritium, percent modern carbon, and age classification of samples, San Francisco Bay study unit, California GAMA Priority Basin Project.

[<, less than; Modern, sample with water recharged since 1952; Pre-modern, sample with water recharged before 1952; Mixed, sample with modern and pre-modern components; na, not available]

USGS GAMA well identification number	Tritium, tritium units	Percent modern carbon (percent)	Age classification	USGS GAMA well identification number	Tritium, tritium units	Percent modern carbon (percent)	Age classification
SF-01	5.3	88	Mixed	SF-41	2.9	na	Modern or mixed
SF-02	3.2	85	Mixed	SF-42	< 1.0	2	Pre-modern
SF-03	2.1	na	Modern or mixed	SF-43	< 1.0	12	Pre-modern
SF-04	< 1.0	66	Pre-modern	SFM-A1	< 1.0	16	Pre-modern
SF-05	< 1.0	51	Pre-modern	SFM-A2	< 1.0	72	Pre-modern
SF-06	2.2	107	Modern	SFM-A3	1.5	66	Mixed
SF-07	5.4	95	Modern	SFM-A4	4.3	87	Mixed
SF-08	1.4	45	Mixed	SFM-B1	< 1.0	41	Pre-modern
SF-09	1.2	82	Mixed	SFM-B2	< 1.0	51	Pre-modern
SF-10	< 1.0	56	Pre-modern	SFM-C1	< 1.0	8	Pre-modern
SF-11	2.4	91	Modern	SFM-C2	2.8	82	Mixed
SF-12	2.9	100	Modern	SFM-C3	2.5	81	Mixed
SF-13	1.4	81	Mixed	SFM-C4	2.5	85	Mixed
SF-14	1.2	77	Mixed	SFM-C5	3.0	86	Mixed
SF-15	< 1.0	51	Pre-modern	SFM-D1	< 1.0	9	Pre-modern
SF-16	< 1.0	43	Pre-modern	SFM-D2	< 1.0	51	Pre-modern
SF-17	2.1	82	Mixed	SFM-D3	3.4	78	Mixed
SF-18	1.8	53	Mixed	SFM-D4	4.2	92	Modern
SF-19	< 1.0	37	Pre-modern	SFM-E1	< 1.0	38	Pre-modern
SF-20	3.2	102	Modern	SFM-E2	3.4	87	Mixed
SF-21	3.1	98	Modern	SFM-E3	31	97	Modern
SF-22	2.7	108	Modern	SFM-F1	< 1.0	5	Pre-modern
SF-23	3.0	79	Mixed	SFM-F2	< 1.0	2	Pre-modern
SF-24	3.1	85	Mixed	SFM-F3	1.3	15	Mixed
SF-25	3.5	75	Mixed	SFM-F4	< 1.0	17	Pre-modern
SF-26	1.5	102	Modern	SFM-F5	< 1.0	32	Pre-modern
SF-27	2.0	92	Modern	SFM-F6	< 1.0	75	Pre-modern
SF-28	3.3	97	Modern	SFU-01	< 1.0	45	Pre-modern
SF-29	2.6	100	Modern	SFU-02	1.6	78	Mixed
SF-30	2.2	87	Mixed	SFU-03	1.8	81	Mixed
SF-31	3.2	99	Modern	SFU-04	1.9	85	Mixed
SF-32	2.7	94	Modern	SFU-05	2.3	87	Mixed
SF-33	1.9	79	Mixed	SFU-06	2.0	69	Mixed
SF-34	4.0	89	Mixed	SFU-07	3.4	90	Mixed
SF-35	6.3	36	Mixed	SFU-08	3.2	95	Modern
SF-36	3.9	102	Modern	SFU-09	6.2	98	Modern
SF-37	3.8	94	Modern	SFU-10	14	88	Mixed
SF-38	< 1.0	24	Pre-modern	SFU-11	< 1.0	93	Mixed
SF-39	< 1.0	24	Pre-modern	SFU-12	na	92	Modern or mixed
SF-40	< 1.0	57	Pre-modern				

Appendix E. Comparison of Data from the CDPH and USGS-GAMA Program

CDPH and USGS-GAMA data were compared to assess the validity of combining data from these different sources. Because the reporting levels for most organic constituents and trace elements were substantially lower for data collected by the USGS GAMA Priority Basin Project than for data from the CDPH database (table 3), only relatively high concentrations of constituents could be compared, and as a result, there were insufficient data to rigorously evaluate agreement between CDPH and USGS-GAMA data. However, concentrations of inorganic constituents (calcium, magnesium, sodium, alkalinity, chloride, sulfate, TDS, and nitrate), which generally are prevalent at concentrations substantially greater than reporting levels, were compared for each well using data from both sources. Six wells had major ion data, and 39 wells had nitrate data from the USGS database and from the CDPH database. Although differences between the paired datasets existed for a few wells, most sample pairs plotted close to a 1:1 line (r^2 = 0.98) (fig. E1). These plots indicated that the GAMA and CDPH inorganic data were comparable.

Major ion data for grid wells with sufficient data (USGS and CDPH data) were plotted on Piper diagrams (Piper, 1944) with all CDPH major ion data to determine whether the grid wells represented the range of groundwater types that have historically been observed in the study unit. Piper diagrams show the relative abundance of major cations and anions (on a charge equivalent basis) as a percentage of the total ion content of the water (fig. E2). Piper diagrams often are used to define groundwater type (Hem, 1985). All cation/anion data in the CDPH database that had a cation/anion balance less than 10 percent were retrieved and plotted on these Piper diagrams for comparison with USGS-grid well data.

The ranges of water types in grid wells and noted historical CDPH data were similar (fig. E2). In most wells, no single cation accounted for more than 60 percent of the total cations, and bicarbonate accounted for more than 60 percent of the total anions. Waters in these wells are described as *mixed cation–bicarbonate* type waters. There were also wells that contained *mixed cation–mixed anion* type waters, indicating that no single cation and no single anion accounted for more than 60 percent of the total, and some wells contained *calcium/ magnesium–bicarbonate* type waters, for which calcium plus magnesium and bicarbonate account for more than 60 percent of the cations and anions, respectively.

The determination that the range of relative abundance of major cations and anions in grid wells is similar to the range of those in all CDPH wells indicates that the grid wells represent the types of water present within primary aquifer system in the SFBAY study unit.

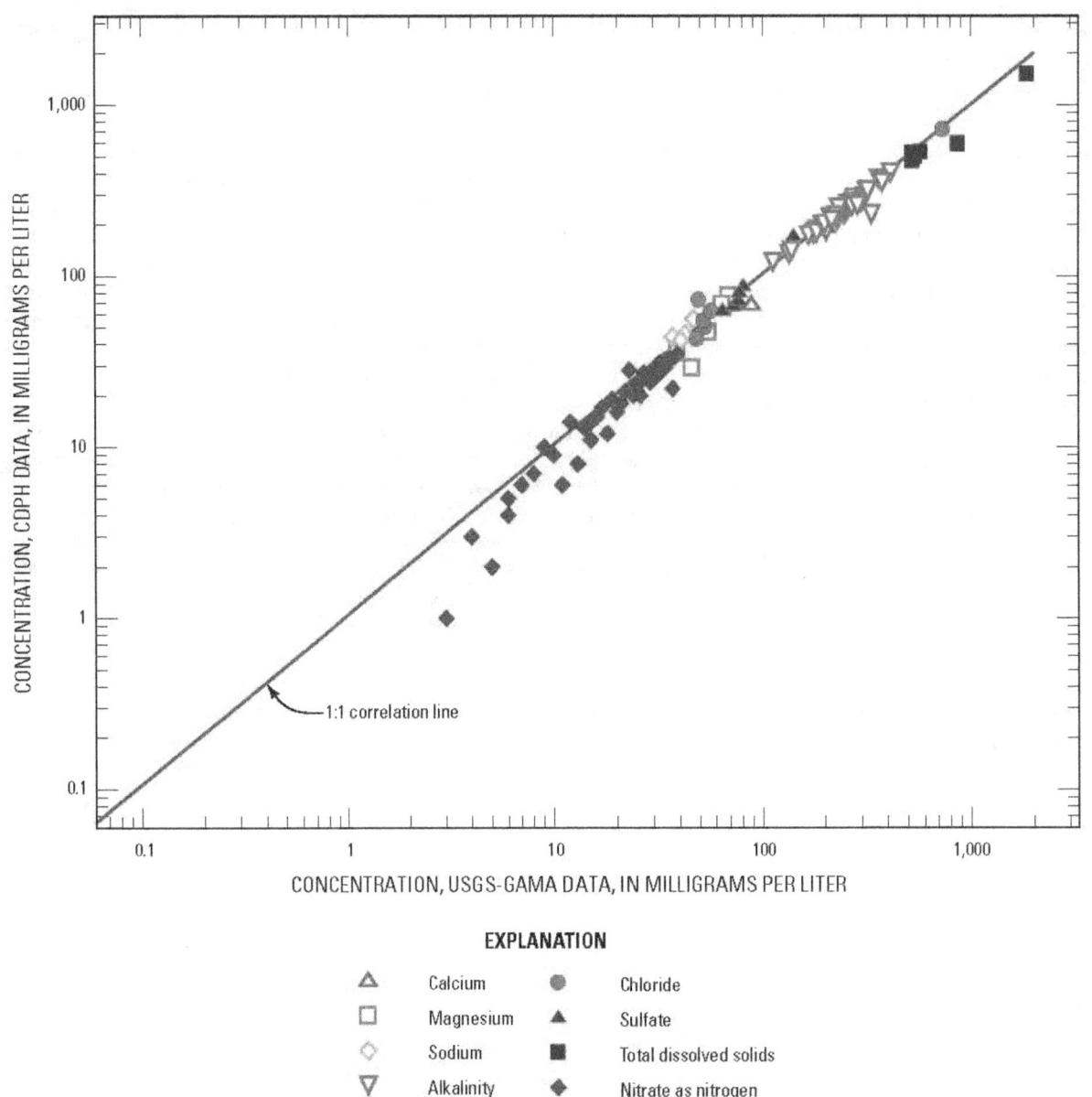

Figure E1.   Paired inorganic constituent concentrations from wells sampled by USGS GAMA Priority Basin Project from April to June 2007, and California Department of Public Health (CDPH) from April 1, 2004, to March 31, 2007, San Francisco Bay study unit, California GAMA Priority Basin Project.

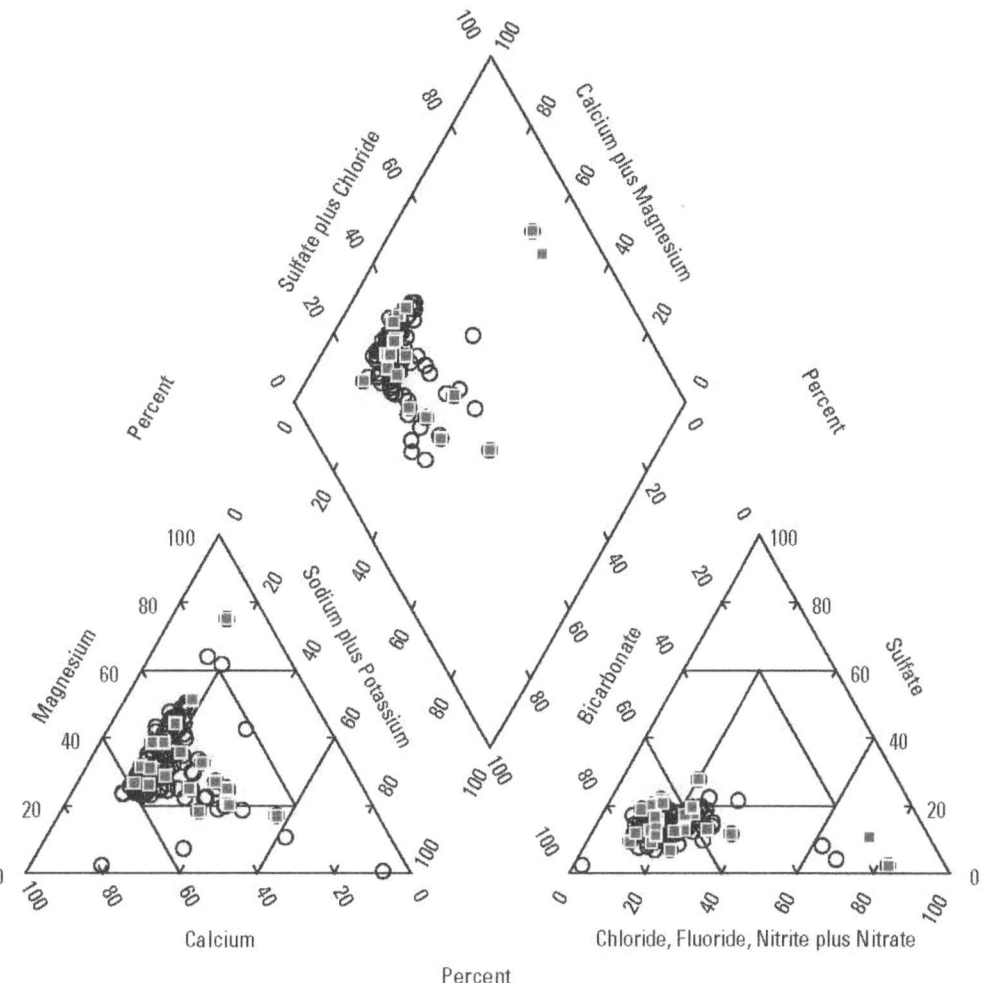

Figure E2.   Selected inorganic data from USGS-grid wells and from all wells in the California Department of Public Health (CDPH) database that have a charge imbalance of less than 10 percent, San Francisco Bay study unit, California GAMA Priority Basin Project.